SpringerBriefs in Systems Biology is an exciting new series of concise publications of cutting-edge research and practical applications in Systems Biology. Systems Biology is the study of the complex interactions between the components of biological systems (genes, proteins, mechanisms, etc), and how these interactions give rise to the function and behavior of that system. The structure and dynamics of cellular and organismal function are examined as a whole, rather than as isolated parts. The interaction of these parts gives rise to new properties and functions which are called "emergent properties".

SpringerBriefs in Systems Biology

David R. Bickel

Phylogenetic Trees and Molecular Evolution

A Hands-on Introduction with Uncertainty
Quantification Corrected

 Springer

David R. Bickel (iD)
Informatics and Analytics
University of North Carolina at Greensboro
Greensboro, NC, USA

ISSN 2193-4746   ISSN 2193-4754   (electronic)
SpringerBriefs in Systems Biology
ISBN 978-3-031-11957-6   ISBN 978-3-031-11958-3   (eBook)
https://doi.org/10.1007/978-3-031-11958-3

This Springer imprint is published by the registered company Springer Nature Switzerland AG
The registered company address is: Gewerbestrasse 11, 6330 Cham, Switzerland

To Mary Anna, beloved and highly favored

μακαρια η πιστευσασα οτι εσται τελειωσις τοις
λελαλημενοις αυτη παρα κυριου

Preface

Why This Book?

Brief Description

The book introduces molecular phylogenetics and explains how to interpret the results of evolutionary models, which by their nature only quantify a fraction of the uncertainty involved. The book is intended to introduce upper-level undergraduate and graduate students of biology to phylogenetic tree reconstruction and its dependence on models of molecular evolution. The scope encompasses basic concepts of estimating divergence times and ancestral sequences and of quantifying the uncertainty of the estimates.

Key features:

- Selective use of mathematics

  - An informal, graphical introduction to the main concepts of molecular phylogenetics
  - A competitive dice game to teach evolution modeling without equations
  - Gradual exposure to simple formulas for understanding phylogenetic methods

- Uncertainty quantification

  - Thorough explanations of sources of uncertainty in phylogenetics
  - Simple methods for quantifying uncertainty not captured by phylogenetics software

- Concepts underlying present and future software more than details of today's programs

Why the Subtitle "A Hands-On Introduction with Uncertainty Quantification Corrected?"

This book serves two purposes. First, it introduces biology students to phylogenetic trees and molecular evolution without requiring any mathematics beyond elementary algebra. It does so by presenting the main concepts in a variety of ways: first in graphics, then in some history of the field, next in a dice game, and finally in simple equations. Completing the exercises will prepare students for textbooks with more realistic mathematical modeling and details of current software.

Second, the book explores the interface between molecular evolution and uncertainty quantification, of potential benefit to both fields. For not only does phylogenetics clearly illustrate the need for uncertainty quantification in science, but quantifying the uncertainty involved in estimating phylogenetic trees can lead to reliable biological interpretations. This book equips students with statistical tools for correcting uncertainty quantification for the study of evolutionary relationships between DNA and protein sequences.

Chapter-by-Chapter Synopsis

1. Chapter 1 starts slow, without assuming more than a basic knowledge of biology or mathematics. Readers are exposed to the main concepts of the book through graphics and an easy-to-read, conversational exposition. The absence of formulas facilitates comprehension for readers intimidated by mathematics. The chapter ends with simple exercises to reinforce the basic ideas.
2. This chapter continues to expose readers to the main concepts of the book. It does so by sketching a history of the main developments of molecular phylogenetics and molecular evolution since the 1960s. All mathematical formulas related to this chapter are postponed until Appendix B and Appendix C. The exercises at the end of the chapter encourage reflection, especially on how the history can inform assessments of the uncertainty in estimated trees.
3. The third chapter covers a probability model of molecular evolution in a way designed to engage readers with little interest in mathematics. That is accomplished by stating the model in terms of a competitive game rather than mathematical formulas: one team rolls dice to simulate molecular evolution, and the other team then uses the resulting simulated sequence data to estimate the phylogenetic tree. The description of the game rules gradually eases the readers into some simple mathematical formulas needed to understand how evolutionary distances are corrected for multiple substitutions. Exercises provided at the end of the chapter give readers hands-on experience with the sequence simulation and tree reconstruction methods.
4. The fourth chapter goes into more detail about estimating divergence times from molecular sequence data. It explains the inadequacy of confidence intervals for

quantifying the uncertainty in divergence times. A simple method of improving the uncertainty quantification is provided. Exercises at the end of the chapter give readers experience with quantifying the uncertainty in divergence times.

5. The fifth chapter explains three ways to infer ancestral sequences from sequence data. After explaining a heuristic approach to build the intuition of the readers, the chapter explains maximum likelihood estimation and Bayesian estimation with some simplifications to keep the formulas minimal. To adjust the results for uncertainty not quantified in the models, the previous chapter's method of uncertainty quantification is adapted. Exercises provided at the end of the chapter give readers practical experience with estimating ancestral sequences and intuition about the uncertainty involved.

6. This chapter lifts the uncertainty quantification of the previous chapters up to the level of molecular evolution hypotheses. A relatively recent rival to the neutral theory of molecular evolution is explained in terms of how its predictions differ. The exercises at the end of the chapter give readers experience with quantifying the extent to which sequence data support one evolutionary hypothesis more than another.

7. The last chapter is a guide to further reading on molecular phylogenetics, uncertainty quantification, and other topics encountered in the book.

The three appendices fill in some mathematical gaps.

Acknowledgments

Molecular evolution models were the subject of my PhD research, which would not have been possible without the mentorship of Bruce J. West. This book started in the form of lecture notes at the Medical College of Georgia. I drafted Appendices B and C at the University of Ottawa. Teaching computational biology courses at the University of North Carolina at Greensboro led to many updates and additions, and I am grateful to all the students who provided feedback.

The comments of the anonymous reviewers on the book proposal led to the addition of Chap. 7 and to an expansion of Sect. 2.7. I thank Shi Huang for carefully reading a draft of Chap. 6 and for suggesting corrections.

Larissa Albright, my first Springer contact for this book, ensured efficient peer review and publishing agreement processes. I appreciate how quickly Sanjana Sundaram and Merry Stuber answered all my questions as I finalized the manuscript for submission to Springer. I thank Tiffany Lu for helpful correspondence and both Sanjana Sundaram and Vinodhini Srinivasan for coordinating the publication process. I am grateful to A. Meenahkumary at Straive for efficiently managing production. The comments of the two anonymous reviewers led to improved clarity and scope.

My interest in dice games has its origins in unforgettable campaigns with Chris, Brian, John, and David. I warmly thank Dorothy Johnson for her sustained interest in the progress of the book project. Mary Anna deserves mention for her professionalism in securing the epigraphs' permissions. I enjoyed discussing the topic of the book with her, Evelyn, and Christian and seeing the reaction of Faith and Lydia to the plots of upside-down trees!

Greensboro, NC, USA   David R. Bickel
May 2022

Contents

Chapter 1
Introduction to Phylogenetic Trees

The most important fact to keep in mind about any phylogenetic tree is that it is not fact, it is an estimate—an inference. . . . An estimate is a fine and useful thing; just keep in mind that it is not absolute truth.

–B. G. Hall

1.1 What Are Phylogenetic Trees All About?

Suppose you want to make a hypothesis about how three DNA virus variants are related. You know their DNA sequences only differ in two locations, called *sites*.

The segment of two sites will be considered here as the sequence relevant to making guesses about past evolution. Then there is a nucleotide (**a**, **c**, **g**, or **t**) at each site of each of the three sequences. Let us say the sequences are **ag**, **ca**, and **ga**. We assume that they come from the same ancestral sequence by a series of changes or *substitutions* of one nucleotide for another. Each substitution resulted from a mutation that spreads to the entire population of the virus variant. Here is a possible way it could have happened:

Electronic Supplementary Material The online version contains supplementary material available at (https://doi.org/10.1007/978-3-031-11958-3_1).

[1] *Phylogenetic Trees Made Easy: A How-To Manual* (Sinauer Associates) [53, p. 3]. © 2018 Oxford University Press. Reproduced with permission of the Licensor through PLSclear.

D. R. Bickel, *Phylogenetic Trees and Molecular Evolution*, SpringerBriefs in Systems Biology, https://doi.org/10.1007/978-3-031-11958-3_1

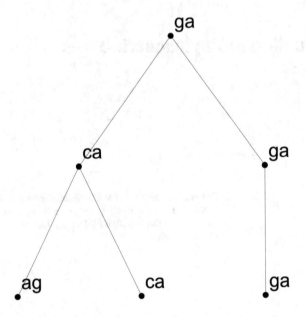

That possible *phylogenetic tree* hypothesizes the following scenario. The top or *root* of the tree is the ancestral sequence, which is **ga** in this case, and the *tips* of the tree on the bottom are the three present-day sequences, the ones we actually know. The left-hand branches of the tree say the top **ga** sequence evolved to the intermediate **ca** sequence, which then evolved to **ag** on the far left but was left unchanged in the branch leading to the **ca** tip. On the right-hand side of the tree, we see that the sequence is not changed. Each branch of the tree represents a different lineage, a different variant of the virus according to its DNA sequence. Here is the same tree with the branches labeled by their numbers of nucleotide differences:

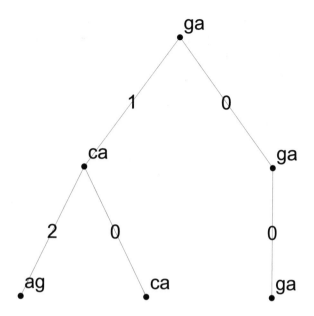

Then the number of nucleotide differences from the *root* or *common ancestor* **ga** to **ag** is $1 + 2$, the number from the root **ga** to the **ca** tip sequence is $1 + 0$, and the number from the root **ga** to the **ca** tip is $0 + 0$. That might be explained if there were evidence that the rate of molecular evolution was fastest on the left and slowest on the right.

How related is the **ag** tip to the **ca** tip? Since there are 2 nucleotide differences between the **ca** ancestor and **ag** and 0 between the **ca** ancestor and the **ca** tip, the answer is $2 + 0$ substitutions. That agrees with the number of nucleotide differences directly observed between the **ag** and **ca** tips, for they differ from each other at both of their sites: **a** differs from **c** at the first site, and **g** differs from **a** at the second site.

How related is each of those tips to the **ga** tip? From the **ca** tip to the **ga** tip, we count on the tree 0 differences from **ca** to **ca** $+ 1$ difference from **ca** to the root **ga** $+ 0$ from the root **ga** to the intermediate **ga** $+ 0$ from the intermediate **ga** to the tip **ga**, for a total of $0 + 1 + 0 + 0 = 1$ difference. That is the same as the observed number of differences between the **ca** and **ga** tips, for **c** differs from **g** at the first site, and those tips agree at the second site, both having **a** there.

Likewise, going dot to dot (node to node) from the **ag** tip to the **ga** tip, there are $2+1+0+0$, for a total of 3 nucleotides according to the tree. But by directly counting the number of differences between **ag** and **ga**, there are only 2 differences: **a** differs from **g** at the first site, and **g** differs from **a** at the second site. In fact, there could not be more than 2 differences observed since the sequence only has 2 sites. That is the problem of *multiple substitutions* or *multiple hits*: more than one substitution could have taken place in evolutionary history for every observed nucleotide difference between the tips, the present-day sequences.

Another type of multiple substitution is hidden in the tree. It can be uncovered by expanding the branch from the most recent common ancestor of the **ag** and **ca** tips to the **ca** tip. That branch is from a **ca** sequence to another **ca** sequence, and yet this is a real possibility:

That says the top **ca** sequence changed at least $1 + 1 + 1$ times, for a total of 3 nucleotide differences on the tree, and yet resulted in a **ca** tip at the bottom, exactly matching the starting sequence. That phenomenon is called *back substitution*, since a site reverts back to the nucleotide, it had been before a previous substitution. That might remind you of when you changed a Wikipedia page that someone else changed back while you were sleeping. At least three nucleotide changes took place, and yet the number of differences is 0 between the **ca** ancestor at the top and the **ca** tip at the bottom. The problem of multiple substitutions can be solved to some extent by using correction formulas that rely on mathematical models of molecular evolution.

Many other possibilities for trees could have resulted in the three tip sequences observed. For instance, if the rate of evolution were faster than we thought, the sequences could have evolved like this:

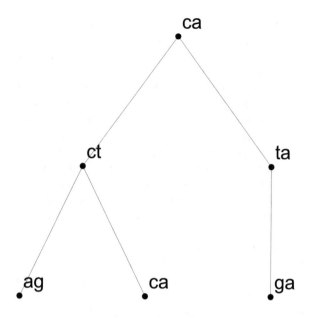

How many nucleotide differences would go on each branch of that tree?

A convenient way to refer to the sequences on any of the above trees is to label them by their relation to the two present-day sequences that are most closely related:

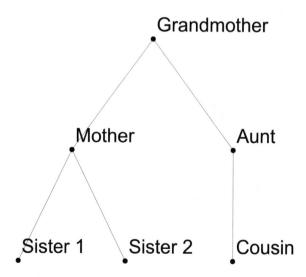

That should not be taken too literally: each branch in the tree represents several generations. Just as you are more closely related to your sibling than you are to your cousin, the sequences labeled "Sister 1" and "Sister 2" are more closely related to each other than to the sequence labeled "Cousin."

The three present-day sequences at the tips give clues not only about the ancestral sequence at the root but also about when the virus variants may have split from each other in a speciation event. When did the Mother sequence start evolving separately from the Aunt sequence? When did the population of Sister 1 get reproductively separated from the population of Sister 2? The numbers of months or years since those splits or divergences occurred are called *divergence times*.

Divergence time estimates are calculated based on assumptions about how the rate of evolution might differ from branch to branch. The simplest and most common calculations are based on rough approximations of the *molecular clock hypothesis*, the idea that the rate of evolution is the same for all branches. The first tree we tried for the viral sequences seems to violate the hypothesis, with the branches on the left evolving faster than those on the right. That tree had **ag** and **ca** as Sister sequences with **ga** as the more distant Cousin.

What if the evolution of the virus sequences instead followed the molecular clock? Noting that **ca** and **ga** only differ at the first nucleotide site and that **ag** has no site in common with them, it is more natural to think of **ca** and **ga** as Sister sequences with **ag** as the more distant Cousin. Then this would be a more plausible way **ag**, **ca**, and **ga** are related:

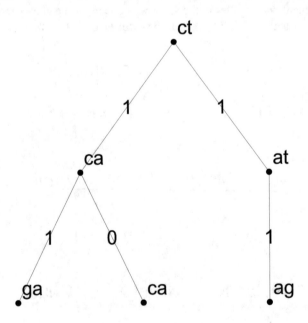

But many other trees that also have **ag** as the Cousin would be just as plausible, including this one:

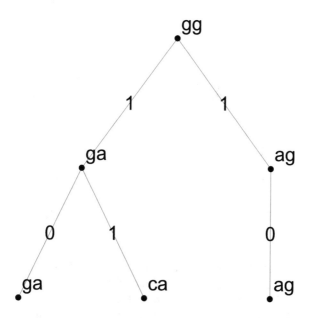

So from our sequence data, while we might be willing to hypothesize that **ag** is the Cousin, there is too much uncertainty about the Grandmother (the common ancestor at the root) to venture a guess about its sequence. Putting numbers to the levels of uncertainty in these kinds of inferences about past events will be covered in some detail, when we get to *uncertainty quantification.*

Say we know not only the three tip sequences but also that the most recent common ancestor was found about 3 years ago. That means it took 3 years for the Grandmother sequence to evolve to the **ga** tip, 3 years to evolve to the **ca** tip, and 3 years to evolve to the **ag** tip. The 3 years is the divergence time, the time for all the lineages leading to the present-day sequences to diverge from each other.

Now we would also like to estimate the divergence time for the most recent common ancestor of the **ga** and **ca** tips, assuming that they are Sisters and that **ag** is their Cousin. To guess how many years passed since the lineages of the **ga** and **ca** tips diverged from each other, we could start averaging the numbers of nucleotide differences of the last two trees without worrying about what the ancestral sequences are. That would give us this tree:

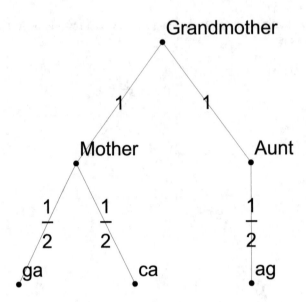

For simplicity, let us assume that there were not any multiple substitutions. In that case, the numbers of nucleotide differences on the averaged tree are our estimates of the numbers of substitutions that occurred. Then, no matter which tip we choose, we guess based on the averaged tree that about $1 + 1/2$ substitutions took place between Grandmother and the tip. For example, it looks like 1 substitution from Grandmother to Aunt, plus half of a substitution from Aunt to the **ag** tip. While the actual number of substitutions cannot be 1/2, that is still our best guess since we do not have enough information to decide between 0 substitution and 1 substitution.

In that way, we estimate that it took 3 years to get about 1.5 substitutions for each of the three lineages. That is a rate of one substitution every two years, which is the same as half of a substitution per year. The rate of 0.5 substitutions/year suggests doubling the numbers of nucleotide substitutions in the averaged tree to get this *time tree*:

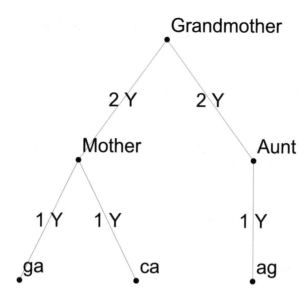

The Y stands for "year" or "years." We say the time tree was *calibrated* by the divergence time of 3 years since the Grandmother split into the Mother and Aunt lineages. We see from the time tree "1 year" between Mother and **ga** and the same amount of time between Mother and **ca**. We conclude that our rough estimate of the **ga-ca** divergence time is 1 year.

Unlike the divergence times of the virus variants, divergence time estimates are in millions of years ago when they are calibrated by the fossil record. Calibrating trees by the fossil record adds its own uncertainties, including both measurement and systematic errors. For example, if species actually diverged millions of years before conditions were suitable for fossilization, then the time used to calibrate the rest of the time tree will be too small by millions of years. Later in the book, we will see a method of quantifying uncertainties of that type as well as others.

But we can already get a feeling for how uncertainties propagate through a tree by returning to our three viral sequences. Instead of averaging our best two candidate trees, let us express some of their uncertainty as the ranges they give the branches that lead immediately to the three tips:

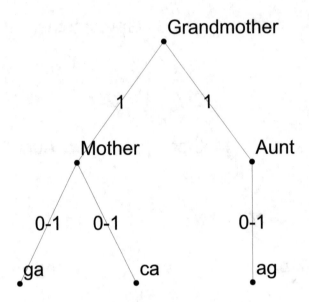

Notice that the new tree is the same as the averaged tree except that it has "0–1" as the nucleotide substitutions instead of each "1/2." For all we know, the number of substitutions on the branches right before the tips could be 0 as easily as 1. Recall that the 3-year divergence time since the Grandmother, the most recent common ancestor, was an approximation. Say it could be anywhere from 2 years to 4 years (2–4 years).

So we are uncertain not only about the tree structure and the branch lengths (numbers of nucleotide differences) but also about the Grandmother divergence time to be used to calibrate the tree into a time tree. How do those uncertainties affect our uncertainty about the Mother divergence time, the number of years, since **ga** and **ca** diverged from their most recent common ancestor?

A cautious way to answer that without making more assumptions is to see what would happen if each of the extremes in the ranges were the true value. The number of nucleotide differences between Grandmother and any of the three tips is between $1 + 0$ and $1 + 1$, for a range of 1–2 differences. Again ignoring the possibility of multiple substitutions, that would mean it took 2–4 years for 1 or 2 substitutions to take place in each lineage. So the rate of evolution between 1 substitution per 4 years (0.25 change per year) at the slowest and 2 substitutions per 2 years (1 change per year) at the fastest. In short, the rate is expressed by the range of 0.25–1 substitutions/year.

To calibrate the tree by that range of rates, we just divide each branch length by each extreme in the range. First, the 1 on the branch from Grandmother to Mother becomes at least (1 substitution)/(1 substitution/year) = 1 year and at most (1 substitution)/(0.25 substitutions/year) = 4 years, so we will write 1–4 years on that branch of the time tree to indicate that as the amount of time between Grandmother and Mother. The same is true of the branch from Grandmother to Aunt since it also shows 1 nucleotide substitution.

The same range of 1–4 years would hold for the branch from Mother to **ga** for the upper part of its 0–1 range displayed as its branch length. The lower part of that range is 0 nucleotide substitutions, which calibrates to at least (0 substitution)/(1 substitution/year) = 0 years and at most (0 substitutions)/(0.25 substitutions/year) = 0 years. The range is 0–0 years or just 0 years regardless of the rate since it takes 0 years for 0 substitutions to occur. Then using 0 years as the least amount of time that elapsed since **ga** diverged from **ca** and 4 years as the longest amount of time, we get 0–4 years as the range of divergence times for the branch from Mother to **ga**. The tree also indicates 0-1 substitutions on the other two final branches leading to **ca** and **ag**, so each of their divergence time ranges is also 0–4 years.

Putting everything together, we get this time tree:

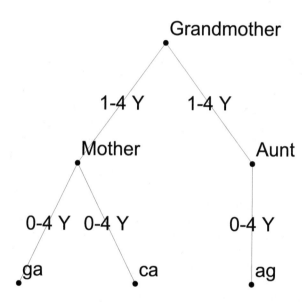

The ranges on the branches reflect the two sources of uncertainty we have considered: the numbers of nucleotide differences and the divergence time used to calibrate the tree. If we had included more sources of uncertainty in the analysis, we would have seen even wider ranges of times. On the other hand, the ranges could be reduced if we knew more about how uncertainty propagates through the tree. While some sources of uncertainty could also be reduced by acquiring more sequence data, that has no effect on other sources of uncertainty (Sect. 4.2.1.2).

1.2   Assumptions Behind Models of Molecular Phylogenetics

Several assumptions about molecular evolution are so commonly made and so necessary to most evolutionary analyses of sequence data that they are seldom stated explicitly. Knowing those assumptions gives an idea of the reliability of studies of molecular evolution that use them a method cannot warrant more confidence than its assumptions. Methods for quantifying the some of the uncertainty in conclusions drawn from uncertain assumptions will be presented in Sects. 4.1, 4.2, and 5.5.

How many uncertain assumptions were already made in the introduction (Sect. 1.1)? This section explains some of them in more detail. More will be covered in Sect. 3.4.

1.2.1   Common Ancestry

Models of molecular evolution and phylogenetic trees constructed from DNA or protein sequences depend on the assumption that the sequences involved share a common ancestor. Since it is the basis for comparative sequence analyses, it cannot be supported by such analyses alone without circular reasoning. Even though scientists almost universally accept that all organisms descended from a single common ancestor (Sect. 6.4), the common ancestry needed for phylogenetic trees is *homology* in the sense that the observed similarity in sequences is due to descent from a common ancestral sequence [53, chapter 3]. While high degrees of sequence similarity provide some evidence of homology, such evidence is inconclusive apart from other evidence (Sect. 3.4.1).

An open question is whether a phylogenetic tree constructed from sequences that are not homologous has useful information even though it fails to accurately estimate an evolutionary tree. Schwabe [105] argued that published phylogenies give structural, but not necessarily evolutionary, relationships. They would then be interpreted in terms of hierarchical cluster analysis, an approach to unsupervised machine learning that is used to group similar entities or features together, keeping dissimilar ones apart, without any expectation of evolutionary relationships (Sect. 7.2). In fact, some phylogenetic tree estimators are considered to be methods of hierarchical cluster analysis [33, 44, 121] even though clusters are not usually interpreted as phylogenetic trees. An estimator of that type is explained in Sect. 3.1.2.1.

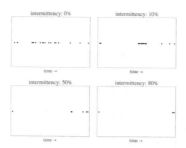

Fig. 1.1 Each of the four plots displays 20 substitution events, which appear to overlap more and more as the intermittency increases. The plot with 0% intermittency was simulated from a Poisson process (see Sect. 1.1). Each of the other plots was simulated from a fractal renewal process [21, 84] with independent inter-event times following a Pareto distribution according to the displayed "intermittency," which is explained in Sect. 2.8 and defined in Appendix B

1.2.2 Molecular Clock Hypothesis

Figure 1.1 displays possible substitutions over time according to four different models of molecular evolution rates. Only the plot with 0% intermittency corresponds to the molecular clock hypothesis of Sect. 1.1; the other three plots deviate from it more and more as the intermittency increases.

Although the molecular clock hypothesis is still used to simplify calculations, sequence comparisons have demonstrated large departures from it, even within each lineage [47, 96]. A more realistic and weaker assumption is stationarity in the sense explained in Sect. 4.2.1. Even when lineage effects are removed, fluctuations in the rate of evolution could have a significant impact on the reliability of inferences about evolution drawn from sequence data [11, 36], so methods that assume a lack of such fluctuations must be used with caution, as will be seen in Sect. 4.2.1.

Nonetheless, the molecular clock hypothesis is a reasonable first approximation whenever the variation of the rate of evolution of a gene between lineages or in time is small compared the variation in the rate between different genes. Indeed, genes can differ from each other in rates of evolution by orders of magnitude.

Modifications of the molecular clock hypothesis underly many current methods of phylogenetic estimation. Those modifications evolved over its history.

1.2.3 Statistical Assumptions

Statistical methods transform sequence data into conclusions, often with a number quantifying the uncertainty of those conclusions. Statistical methods can only do so by relying on assumptions made in the statistical models—assumptions that are themselves uncertain.

The probabilistic statistical methods most commonly used in tree estimation fall under two major categories:

(1) **Frequentist methods** are so-called for their interpretation of probability in terms of hypothetical frequencies of events. These three complementary frequentist methods are often used together in molecular phylogenetics:

   (a) **Confidence intervals** will be introduced in Sect. 4.1.
   (b) **Bootstrapping** is also postponed until Sect. 4.1.
   (c) **Maximum likelihood estimation** will be explained using mathematical notation in Sect. 5.4.

(2) **Bayesian methods** are so-called for their reliance on Bayes's theorem, which transforms data and input probabilities called *prior probabilities* into output probabilities called *posterior probabilities*. The math will be covered in Sect. 5.3.

The two major categories, while at war [89], are increasingly used together. For example, they complement each other in empirical Bayes methods (Sect. 7.2). More generally, frequentist methods, Bayesian methods, empirical Bayes methods, and probabilistic machine learning methods fall under the evidential framework of Appendix A.

1.3 Exercises

(1) Make a glossary of the terms defined in this chapter, writing the definitions in your own words. Then make a plain language glossary defining the same terms in a way non-biologist readers can understand.
(2) What does this chapter say about quantifying uncertainty about phylogenetic trees? What questions about such uncertainty quantification are left unanswered?

Chapter 2
Adaptation of the Molecular Clock: A Divergence Time Story

> *... those who preceded us thought that they too had definitive solutions. We laugh at others and we don't realize that someone will be just as justified in laughing at us on some not too remote day.*
>
> – Nassim Nicholas Taleb[1]

Scientific activities of the early twentieth century that lead to currently accepted theories of molecular evolution include comparisons of molecular differences between species [90] and the neo-Darwinian synthesis of the nineteenth century theories of natural selection and population genetics. This chapter starts with an account of how the molecular clock hypothesis (Sect. 1.2.2) contributed to refuting the neo-Darwinian idea that molecular evolution is mostly driven by the natural selection of beneficial mutations. The rest of the chapter sketches a decade-by-decade history of the molecular clock's mutation to increasingly relaxed forms. Each decade is labeled by one of its main themes.

2.1 1960s: Starting the Clock

Zuckerkandl and Pauline [137] observed that the number of amino acid differences between cyclostome hemoglobin and the hemoglobin of each of the other vertebrate classes is about equal, rather than being less between the cyclostome and the fish than between the cyclostome and the mammal. That observation is called the *genetic equidistance phenomenon* (Sect. 6.2). Margoliash [88] independently

---

Electronic Supplementary Material The online version contains supplementary material available at (https://doi.org/10.1007/978-3-031-11958-3_2).

---

[1] *The Black Swan: The Impact of the Highly Improbable* (Random House) [112].

D. R. Bickel, *Phylogenetic Trees and Molecular Evolution*, SpringerBriefs in Systems Biology, https://doi.org/10.1007/978-3-031-11958-3_2

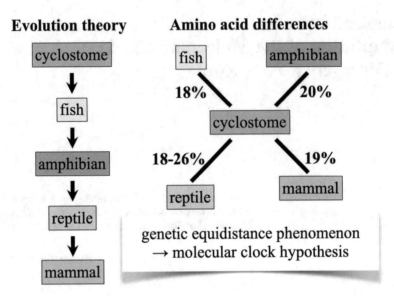

Fig. 2.1 The displayed cytochrome c distances are from lamprey to tuna (18%), bullfrog (20%), turtle (18%), rattlesnake (26%), and human (19%) according to Dayhoff [38, vol. 5, p. D-8]

made the same observation for cytochrome c [97], a phenomenon soon seen more generally. While it at first appears to contradict the evolution of morphology in the sequence cyclostome → fish → amphibian → reptile → mammal (Fig. 2.1), the observation would be compatible with that sequence if the number of substitutions that took place in the lineage from the cyclostome to the modern fish would be about the same as the number in the lineage from the cyclostome to the modern mammal, making changes in protein and DNA approximately independent of morphological change [128].

Zuckerkandl and Pauline [137] took that interpretation of the genetic equidistance phenomenon, concluding that the rate of hemoglobin amino acid substitutions is approximately the same in the lineages studied. They hypothesized the molecular clock in the form of a *Poisson process*, a mathematical model with events as regular as those of radioactive decay. It predicts that the divergence time estimated from the fossil record would be approximately proportional to the estimated numbers of amino acid substitutions. According to the molecular clock hypothesis, the substitution rate for a given protein is constant within a clade or group of species.

While the potential value of the molecular clock hypothesis for estimating divergence times was recognized immediately [138], it ignited a debate about whether most substitutions were due to natural selection or whether they were selectively neutral. Some doubted the constancy of the rate of molecular evolution since the rate of morphological evolution could not possibly be constant [30, 107, discussion by E. Mayr]; environmental changes and natural selection were thought to bring about changes in the rate of evolution. Simpson [107] held that most molecular changes resulted from natural selection.

Indeed, under the neo-Darwinian synthesis, the theory of natural selection accepted at that time, most changes were due to selective pressures, predicting that the rate of evolution would depend on the selective pressures and thus would vary greatly between lineages, contrary to the molecular clock hypothesis. Due to the widespread belief that mutations would only become fixed in populations if they were advantageous, it was surprising to find as much genetic variation as discovered in Drosophila [55, 83].

Kimura [67] and King and Jukes [68] explained the apparent rate constancy by suggesting what became known as the *neutral theory*, according to which most amino acid and nucleotide substitutions are selectively neutral in the sense that they are not subject to natural selection since they do not significantly affect the fitness of the population. Observations consistent with the molecular clock hypothesis were thought to support the neutral theory.

Variations of the neutral theory and the molecular clock hypothesis have been debated since.

2.2   1970s: Neutralism Versus Selectionism

The neutral theory predicted that the rate of substitutions (in populations) was equal to the rate of mutations (in individuals). However, the former rate is constant per year according to the molecular clock hypothesis, and yet the latter seemed to be constant per generation. Some studies around the turn of the century indicating that the rate of amino acid substitution was constant per generation rather than per year were discredited by Wilson et al. [130]. On the other hand, non-coding DNA was found to have a much stronger generation-time effect than coding DNA [97]. Different predictions about generation-time effects were made by the original neutral theory and the *nearly neutral theory* [95], the hypothesis that many molecular substitutions are slightly deleterious rather than completely neutral [97]. The prediction of the neutral theory that the functional significance of a gene correlates with a lower substitution rate has been widely observed [133, p. 392].

On another front, traditional neo-Darwinian models of molecular evolution by natural selection were still advanced in the 1970s [73], supported to some extent by new statistical tests that indicated deviations from the molecular clock hypothesis.

2.3   1980s: Fluctuating Rates

Controversy continued over whether the neutral theory predicted a constant rate per year or per generation. A possible generation-time effect was used to explain reports in 1985 and 1986 that primates evolve slower than rodents and that humans evolve slower than other primates [133, pp. 362]. That idea, called the *hominid slowdown hypothesis*, had already been proposed and refuted in the 1960s [73, p. 655].

The index of dispersion, a measure of deviation of numbers of substitutions from a Poisson process (see Sect. 2.1), was used to detect rate differences between lineages, apparently supporting molecular evolution by natural selection [48, 49] rather than the nearly neutral theory [97]. Takahata [110], however, found that the index of dispersion estimates was consistent with a model of neutral evolution.

The index of dispersion appears to be obsolete now that computer implementations of mathematical models allow the use of tests based on *maximum likelihood estimation*, which finds the values of trees and other model parameters that give the data the highest probability [133, pp. 365–366]. Felsenstein [45] applied maximum likelihood estimation to molecular evolution, providing a heuristic approach to computing the maximum likelihood estimates [43, §14.7].

A nucleotide substitution in a coding sequence is called *synonymous* if, according to the genetic code, it does not lead to an amino acid substitution but is called *nonsynonymous* if it does lead to an amino acid substitution. Differences in estimated rates of evolution between synonymous and nonsynonymous substitutions have been used in the 1980s to estimate the degree of selective pressure in coding sequences [133, p. 390].

2.4   1990s: Confronting the Fossil Record

The nearly neutral theory has proven to be difficult to test [133, pp. 392–393]. Nonetheless, its claim that most nearly neutral substitutions are deleterious was falsified by Gillespie [50] and dropped from the theory [97]. Kuhner and Felsenstein [72] observed that reconstructed phylogenetic trees are sensitive to the assumption that the rate of evolution is constant in time and the same for all branches and to the assumption that the probability of a substitution is independent and identically distributed across nucleotide sites [43, §14.7].

Studies in the 1990s involving more genes than those of the previous decade yielded new results related to the generation-time effect, though still with a hominid slowdown for intron regions [92, p. 190]. The invention of high-throughput sequencing led to much more DNA data for testing hypothesis related to the molecular clock, leading to estimating degrees of differences in rates and explaining those differences rather than testing a universal molecular clock hypothesis [92, pp. 657–658].

The discrepancy between molecular and fossil divergence times came to light in the 1990s using the newly available sequences [73]. Disagreements between molecular biologists and paleontologists erupted from differences in divergence times estimated from sequence data and from the fossil record [40]. For example, the Cambrian explosion of metazoan phyla appears in the fossil record an estimated 500–600 million years ago, but estimates of divergence time based on the molecular clock hypothesis have been up to about twice as old and thus Precambrian. That is one of the disagreements remaining unresolved today (Sect. 4.2.3). In the face of such apparent discrepancies between the fossil record and sequence data, some scientists resorted to the unfashionable hypothesis that life here does not have a common ancestor on earth but rather several microbial ancestors that arrived from space [57]; see Sect. 6.4.

To defend the use of sequence data for estimating divergence times in spite of the fossil record, Donoghue and Smith [40, pp. 29–32] answered various objections. They argued that since the failure of a statistical test to reject the hypothesis of no difference in rates of evolution does not mean there are no differences in rates, allowing differences in rates can yield estimates of divergence times more in agreement with the fossil record. Donoghue and Smith [40, pp. 29–32] also pointed out that some paleontologists admitted that long divergence time estimates could be correct even though not corroborated by the fossil record [cf. 73].

The weakening of the molecular clock hypothesis in the 1990s resulted not only from more data but also from more efficient computer processing. Maximum likelihood methods were used to estimate divergence times under the model of a local clock, assuming that the rate of evolution is constant in some branches of a phylogenetic tree but not for the entire tree [133, §10.3]. A phylogenetic tree constructed by eliminating the branches evolving at different substitution rates is called a *linearized tree* [92, p. 203]. Since 1989, various local clock methods have been used, including a stochastic-rate method [73, p. 659].

Molecular versions of the fossil-based *punctuated equilibrium hypothesis* of rapid evolution at speciation events [51, 52, 93] were entertained. Nichol et al. [93] interpreted evidence for a changing rate of evolution in RNA viruses as support for molecular evolution due to the selection of beneficial mutations, contrary to predictions of the neutral theory. Analyses of mammalian sequence data led to scale-free models of molecular evolution in agreement with punctuated equilibrium [20, 21, 128]. One of those models was used to simulate the data of Fig. 1.1. They are explained in Sect. 2.8.

2.5   2000s: Molecular Punctuated Equilibrium

Bayesian estimates of divergence times were used instead of frequentist methods (Sect. 1.2.3) to incorporate uncertainty and fossil calibrations and calibrations based on estimated times of geological events [133, §10.4]. However, the Bayesian estimates could only integrate multiple sources of uncertainty by accepting uncertain assumptions about prior distributions [28]; see Sect. 4.2.1.5.

With enough genes, the exact molecular clock hypothesis can always be rejected by statistical tests. Nonetheless, the rate of molecular evolution seemed very constant compared to the rate of morphological evolution, leading to wide acceptance of the neutral theory [73, pp. 656–657]. At the same time, objections continued to surface.

For example, a debate in *Science* on the rate of molecular evolution involves a claim of evidence for episodes of many substitutions around the time of speciation events (125, 131, 29, cf. 99). That is the pattern suggested by the fossil-inspired punctuated equilibrium hypothesis mentioned in Sect. 2.4. Webster et al. [125] reported a positive correlation between the amount of genetic change and the net rate of speciation in 30-50% of 56 published phylogenetic trees, where the *net speciation rate* is defined as the speciation rate minus the extinction rate. They saw this correlation as evidence of a molecular punctuated equilibrium corresponding to morphological punctuated equilibrium observed in the fossil record (see Sect. 2.4).

Webster et al. [125] fit each tree by fitting the parameters to maximize the likelihood of the data. Webster et al. [125] used the Kolmogorov–Smirnov test to control the overall error rate even though multiple hypotheses were tested. (On modern empirical Bayes methods of multiple testing, see Sect. 7.2.) A major concern of Witt and Brumfield [131] was that the phylogenetic trees used by Webster et al. [125] lack independence, weighing against that use of the Kolmogorov–Smirnov test. The main criticism that Witt and Brumfield [131] and Brower [29] had in common was that a node in a phylogenetic tree is not necessarily a speciation event.

2.6   2010s: Integration of Data

By this decade, the term "molecular clock" referred not as much to the original molecular clock hypothesis of Sect. 2.1 as to models for inferring divergence times, models that allow the rate of evolution to change in various ways [79]. The observation that divergence time estimates from molecular sequences are often much older than indicated by the fossil record (Sect. 2.4) might be explained if the rate of evolution were much higher around the time of speciation than before and after [79]. That would be a version of the molecular punctuated equilibrium encountered in Sect. 2.5.

For example, while there is still no generally accepted explanation of why the fossil record disagrees with the molecular dating of the divergences of animal phyla (133, p. 362; 26, pp. 458–460), assuming a probability distribution heavily influenced by the fossil record can force estimated rates to be higher at about the time of the Cambrian explosion [80].

Since about 2010, a method called *total-evidence dating* has been used to integrate not only fossil and molecular sequence data but also morphological data [4]. There is also a method for integrating environmental data with fossil and molecular data [34]. Regardless of the method used to combine different types of data, its results must be interpreted cautiously given the sources of uncertainty that its underlying model leaves unquantified (Sect. 4.2.1).

As more data became available, the neutral theory's hold continued to weaken. Kern and Hahn [66] contended that it needs to be replaced by a theory of molecular evolution that explains widespread genomic evidence for the natural selection of beneficial substitutions. An alternative theory motivated by different concerns is examined in Chap. 6.

2.7 2020s: Time Will Tell

Further departures from the molecular clock hypothesis and the neutral theory are expected in the 2020s. Manceau et al. [85] used a new model and analyses of sequence data to argue for a molecular version of punctuated equilibrium (see Sects. 2.4–2.6). Phillips [100] advanced an unconventional analysis of dynein physical interactions, arguing for molecular evolution by natural selection rather than by neutral mutations. Tay et al. [115] studied the COVID-19 impact of an accelerating substitution rate in SARS-CoV-2.

This decade could also see breakthroughs in phylogenomics [23, 32, 39, 117] or phylodiversity analyses [3].

2.8 Excursus: Models with Molecular Evolution over All Time Scales

2.8.1 Why Models with Evolution over All Time Scales?

First, a few definitions. An evolutionary substitution of one nucleotide for another in a coding sequence is called *synonymous* if it does not cause an amino acid substitution according to the genetic code. Otherwise, it is called *nonsynonymous* since it does lead to an amino acid substitution. The *index of dispersion* is a measure of how variable the rate of substitution is.

Regardless of whether synonymous or nonsynonymous substitutions are considered, there is a pattern of correlation between the estimated mean rate of nucleotide substitution and the index of dispersion called a *geometric correlation*. While that geometric correlation is not predicted by most models that modify the molecular clock hypothesis, it is predicted by models of evolution over all time scales [21], called "scale-free models" in Sect. 2.4. If such models might accurately describe changes in the rate of evolution, then conclusions drawn from scale-dependent models may become questionable (Sect. 4.2.1.2).

Examples of scale-free models from the molecular evolution and physics literature are explained next.

2.8.2 Fractal point process models

The proposed *fractal renewal process* may be the simplest model of substitution with molecular evolution on all time scales. Like the constant-rate Poisson process of the molecular clock hypothesis, its times between substitution events are mutually independent, albeit with much higher probabilities of extremely long times. [21].

Models with similar properties are called *fractal point processes* [84], not all of which are fractal renewal processes [10]. For such models, a simple scaling exponent of a fractal point process is interpreted in terms of an *intermittency parameter* that quantifies how episodic events are [10]. The proposed method of quantifying the intermittency of a point process is a simple way to measure how much a fractal point process differs from a Poisson process. Examples were encountered in Fig. 1.1. A method of estimating the intermittency parameter is defined and illustrated by Bickel [10]. Appendix B offers a taste of the mathematics.

2.8.3   Fractal-Rate Poisson Models

2.8.3.1   Generic Fractal-Rate Poisson Model

The *fractal-rate Poisson models* of molecular evolution, like other stochastic-rate Poisson processes, generalize the molecular clock hypothesis by allowing its underlying substitution rate to change over time. However, unlike other stochastic-rate Poisson processes, the fractal Poisson models say molecular evolution takes place across all time scales [20]. Examples include Poisson processes with their substitution rates fluctuating according to normal distributions (Sect. 2.8.3.2) and those that are more episodic (Sect. 2.8.3.3).

2.8.3.2   Gaussian Models of Rates

Bickel [11] explores the impact of variable rates of nucleotide substitution on two statistical methods used in studies of molecular evolution: (1) a test of the molecular clock hypothesis and (2) confidence intervals of numbers of substitutions. Bickel [11] argues that the latter is impacted much more than the former. The main implication would be that variability in the rate of evolution cannot be neglected as a source of uncertainty in the branch lengths of phylogenetic trees [11].

The proposed fractal Gaussian model of molecular evolution is one way to describe the substitution rate as occurring over all time scales. Unlike other models with that feature, its numbers of substitutions follow a normal distribution, which simplified the computation of the confidence intervals of the branch lengths in phylogenetic trees. Such branch lengths represent numbers of nucleotide substitutions.

2.8.3.3   Diffusion Models of Rates

The following diffusion models say molecular evolution is episodic enough to agree with the theory of punctuated equilibrium [128]. While that theory was motivated by the fossil record, the models of molecular evolution were motivated by DNA sequence data.

A Lévy-stable model of molecular evolution differs from the fractal model of Sect. 2.8.3.2 in that its rate agrees with the infinite-variance generalization of the central limit theorem [127]. Models of this type are described mathematically in Appendix C. The fractional-difference model of molecular evolution is also a diffusion model but is instead based on a mathematical device called a shift operator [128].

2.8.4   Multiplicative Model of Molecular Evolution

The rate of DNA evolution considered is the number of nucleotide substitutions per unit of time. Estimates of the substitution rate based on mammalian DNA data agree with the lognormal distributions much better than normal distributions. That is consistent with models of molecular evolution across all time scales [22].

According to the proposed *multiplicative evolution statistical hypothesis*, fluctuations in the substitution rate result from interconnected influences, each of which has a high impact. That is very different from an additive evolution statistical hypothesis, which would instead imply that each influence on the rate has little impact in itself [22].

2.9   Exercises

(1) What did you find surprising about the history of the molecular clock hypothesis?
(2) In your opinion, what were the three most important developments in the history of the molecular clock hypothesis?
(3) How does the history of the molecular clock hypothesis shed light on how much uncertainty there is in phylogenetic trees? Hint: consider the epigraph of this chapter.

2.10   Bibliographic Notes

The account in Sect. 2.1 is largely based on those of Nei and Kumar [92, §10.1] and Yang [133, pp. 360–362]. The account in Sect. 2.2 is indebted to Nei and Kumar [92, pp. 188–189].

Chapter 3
Estimating Phylogenetic Trees

What is dismembered and torn in pieces through extreme
analysis is to be linked again in thought, in a total context of life
that cannot be gathered through construction or reconstruction,
not in intuition or in an irrational way, but rhetorically
arranged, in the medium of history, philosophy, and poetry, and
in the logic of the image and of thought.

– Johann Georg Hamann[1]

3.1 Substituter: The Poisson Game (3 Sequences of 12 Nucleotides Each)

Substituter is a competitive game between two teams. Each team plays in two modes: a simulation mode and an estimation mode. The data set generated by each team in simulation mode is analyzed by the opposing team in estimation mode.

Exercise 1 explains how to start and finish the game. The rules of the game are divided here into one subsection for each mode.

Electronic Supplementary Material The online version contains supplementary material available at (https://doi.org/10.1007/978-3-031-11958-3_3).

[1] As quoted by Oswald Bayer, *A Contemporary in Dissent: Johann Georg Hamann as Radical Enlightener* (Wm. B. Eerdmans Publishing) [9].

D. R. Bickel, *Phylogenetic Trees and Molecular Evolution*, SpringerBriefs in Systems Biology, https://doi.org/10.1007/978-3-031-11958-3_3

3.1.1 Rules for Simulation Mode

3.1.1.1 Simulating Sequences

When in simulation mode, each team rolls either real dice or virtual dice like those at http://bit.ly/2TXZwO5 (accessed August 17, 2021).

Each team follows these steps for *each* expected tree in Fig. 3.1:

(1) Copy the expected tree with its expected branch lengths. For example, when working on expected tree A, copy the three vertical lines displayed in the upper-left image of Fig. 3.1, and write "2" beside each vertical line.
(2) Determine the DNA sequence of each root (ancestor at the top of the tree) by rolling a 4-sided die for each of the 12 nucleotides. See Table 3.1. For example, when working on expected tree A, do that three times since three roots are displayed in the upper-left image of Fig. 3.1. Each of the other three expected trees only has a single root. Write each root's DNA sequence just above the root on the drawing you have from Step 1.
(3) Repeat these steps for each branch of the expected tree, starting from the root (or roots) and ending with the tip sequences at the bottom of the tree:

(a) Roll a 20-sided die to determine the number of nucleotide substitutions according to the branch length given on the expected tree and Table 3.2. (That expected branch length is the *expectation value* of the number of

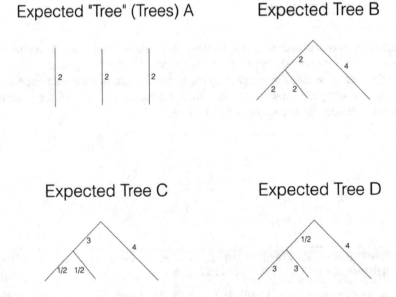

Fig. 3.1 Four expected trees. What is called "Tree A" is actually a group of three trees since its ancestral sequences are not homologous (Sect. 3.4.1). The numbers are expected branch lengths

Table 3.1 4-sided die to simulate the sequence of the top ancestors at the roots of the trees

Outcome of a 4-sided die	Nucleotide
1	a Adenine
2	g Guanine
3	c Cytosine
4	t Thymine

Table 3.2 20-sided die to simulate branch lengths (how many nucleotide substitutions there are in a branch of the realized tree). The first column is the outcome of a 20-sided die. The headings of the other columns are the expected branch lengths (how many nucleotide substitutions there are in a branch of the expected tree)

20-sided die roll	1/4	1/2	3/4	1	2	3	4
1	0	0	0	0	0	0	1
2	0	0	0	0	0	1	1
3	0	0	0	0	0	1	2
4	0	0	0	0	1	1	2
5	0	0	0	0	1	2	2
6	0	0	0	0	1	2	3
7	0	0	0	0	1	2	3
8	0	0	0	1	1	2	3
9	0	0	0	1	2	3	3
10	0	0	1	1	2	3	4
11	0	0	1	1	2	3	4
12	0	0	1	1	2	3	4
13	0	1	1	1	2	3	4
14	0	1	1	1	2	4	5
15	0	1	1	1	3	4	5
16	0	1	1	2	3	4	5
17	1	1	1	2	3	5	6
18	1	1	2	2	4	5	6
19	1	2	2	3	4	6	7
20	2	2	3	3	5	7	8

nucleotide substitutions. It is close to what you would get if you rolled the die 100 times and averaged the resulting numbers of substitutions. Your actual die roll represents the branch length of the *realized tree*; see Nei and Kumar [92, p. 78].)

(b) Write the number of substitutions in parentheses next to the expected branch length.

(c) Repeat these steps for each of those nucleotide substitutions:

   (i) Roll the 12-sided die to determine which of the nucleotides changes.
   (ii) Roll the 6-sided die to determine the base to which it changes according to Table 3.3.

(d) Write the resulting sequence of 12 nucleotides at the node at the bottom of the branch on your drawing of the expected tree.

Table 3.3 6-sided die to simulate substitutions, changes from an old nucleotide to a new nucleotide

Old: a Adenine	Old: g Guanine	Old: c Cytosine	Old: t Thymine	New nucleotide
N/A	1–2	1–2	1–2	a Adenine
1–2	N/A	3–4	3–4	g Guanine
3–4	3–4	N/A	5–6	c Cytosine
5–6	5–6	5–6	N/A	t Thymine

- For example, if there are 0 substitutions according to Step 3b, then the resulting sequence is the same as the initial sequence at the top of the branch.
- Otherwise, if there is 1 substitution according to Step 3b, then the resulting sequence is the same except for the nucleotide that changed according to Step 3c; if there are 2 substitutions according to Step 3b, then the resulting sequence is the same except for the two nucleotides that changed according to Step 3c; etc.

(4) For each tip sequence at the bottom of the tree, determine the protein sequence according to the standard genetic code, assuming that the DNA sequence has 4 codons and thus 4 amino acids (for simplicity, treating the stop codon as if it encoded an amino acid). The standard genetic code is readily available (e.g., [53], Box 1.4 on p. 13; full names of the amino acids in Box 1.3 on p. 10).

- Using a single letter to represent each amino acid will result in a 4-letter protein sequence at each tip of the tree (not at every node), for a total of 3 protein sequences since the expected tree has 3 tips.

3.1.1.2 Sharing Selected Sequences with the Opposing Team

Next, select one of the trees without letting the opposing team know which tree it is. To prevent the opposing team from guessing which tree you choose, you may select it randomly, perhaps by rolling a 4-sided die (1 = Tree A; 2 = Tree B; 3 = Tree C; 4 = Tree D). Follow these steps for the selected tree:

(1) Tell the opposing team the protein sequences, so it may estimate the selected tree according to Sect. 3.1.2.2.

(2) Tell the opposing team the DNA sequences, so it may estimate the selected tree according to Sect. 3.1.2.3.

3.1.2 Rules for Estimation Mode

Section 1.1 introduced the problem of multiple substitutions. If a method of correcting that problem is used in estimating a tree, the tree is called *corrected*; otherwise, it is called *uncorrected*.

With that in mind, you will use the sequences the opposing team gives you (Sect. 3.1.1.2) to estimate phylogenetic trees by following these steps:

(1) For each data set of three **protein** sequences, use the method of Sect. 3.1.2.1 as directed by Sect. 3.1.2.2.

- Result: an uncorrected protein-sequence tree and a corrected protein-sequence tree

(2) For each data set of three **DNA** sequences, use the method of Sect. 3.1.2.1 as directed by Sect. 3.1.2.3.

- Result: an uncorrected DNA-sequence tree and a corrected DNA-sequence tree

3.1.2.1 Distance-Based Tree Estimation

A simple way to estimate a tree from a protein or DNA sequence is based on the concept of an evolutionary distance between any two sequences. The distance could be defined as the number of differences between the two sequences; other definitions of distance will be defined in Sects. 3.1.2.2 and 3.1.2.3.

If there are three available sequences, as in Sect. 3.1.1.1, then this method, displayed in Fig. 3.2, can then be used to estimate a tree:

(1) Label the three sequences by x, y, and z.
(2) Write down the *distance matrix* given in Table 3.4 with the distance between x and y in place of $\overline{xy}$, the distance between x and z in place of $\overline{xz}$, and the distance between y and z in place of $\overline{yz}$. The remaining steps refer to those three distances as the *calculated distances* to distinguish them from the zeros in the diagonal.
(3) Are two of the three calculated distances equal to each other and less than the other calculated distance? In other words, $\overline{xy} = \overline{xz} < \overline{yz}$, $\overline{xy} = \overline{yz} < \overline{xz}$, or $\overline{yz} = \overline{xz} < \overline{xy}$?

    (a) If so, then it appears that the rate of evolution changed too much to apply this method of estimating a tree. That means it cannot be used since it depends on the assumption of the *molecular clock hypothesis*, which says the rate of evolution did not change (cf. Sect. 6.2). In this case, write "molecular clock N/A" and start over at Step 1 using a different data set of three sequences generated by the opposing team.

    (b) Otherwise, proceed to the next step.

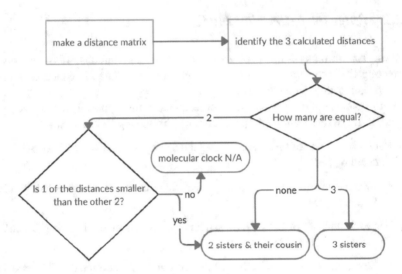

Fig. 3.2 Flowchart for distance-based tree estimation

Table 3.4 Distance matrix used by Step 2 of Sect. 3.1.2.1. Each row and column label is a molecular sequence. Each cell represents a distance between protein sequences (Sect. 3.1.2.2) or DNA sequences (Sect. 3.1.2.3). The diagonal is full of zeros since each sequence is not at all different from itself

	x	y	z
x	0	$\overline{xy}$	$\overline{xz}$
y	$\overline{xy}$	0	$\overline{yz}$
z	$\overline{xz}$	$\overline{yz}$	0

(4) If the three calculated distances are all equal to one another ($\overline{xy} = \overline{xz} = \overline{yz}$), then copy Fig. 3.3 as the estimated tree, and label each branch of the tree with the number equal to $\overline{xy}/2$ as the estimated branch length, again substituting the distance between x and y for $\overline{xy}$. If $\overline{xy} = \overline{xz} = \overline{yz}$, you have then completed the tree estimation. If not, then proceed to the next step.

(5) Check to make sure that one of the calculated distances is less than the other two calculated distances. If there is not a unique shortest distance, then you made a mistake and need to return to Step 3. Otherwise, complete the tree estimation by following these steps:

(a) Copy Fig. 3.4 as the estimated tree.

(b) On that estimated tree, label each of the sisters with one of the sequences involved in the shortest distance, and label the cousin with the other sequence. For example, if $\overline{yz} < \overline{xz}$ and $\overline{yz} < \overline{xy}$, then you would write x by "Cousin," and you could write y by "Sister 1" and z by "Sister 2." The sisters are considered to be in the same cluster or *clade*, which does not include the cousin.

Fig. 3.3 The estimated tree under the conditions of Step 4 of Sect. 3.1.2.1. The occurrence of more than two direct descendants from an ancestor is called a *polytomy*

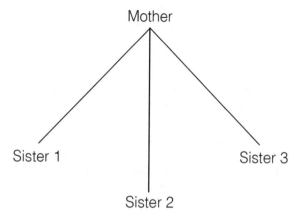

Fig. 3.4 The estimated tree under the conditions of Step 5 of Sect. 3.1.2.1. Since each ancestor has exactly two direct descendants, the tree is called *bifurcating*

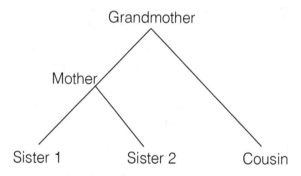

(c) The unique shortest distance among the three calculated distances is the distance between the two sisters on the tree. For that reason, on the branches of the tree between the mother and each of the sisters, write down the number equal to half of that distance as each of the estimated branch lengths. In the example of Step 5b, that would be $\overline{yz}/2$.

(d) On the branch of the tree between the hypothetical common ancestor ("Grandmother") and the cousin, write down the number equal to half of the average distance from the cousin to each of the sisters. That number is equal to 25% of the sum of the two distances from the cousin to the sisters. In the example of Step 5b, that would be $(\overline{xz} + \overline{xy})/4$.

(e) The only branch without an estimated branch length written on it is the branch between the hypothetical common ancestor ("Grandmother") and the mother. For it, write down the branch length calculated in Step 5d minus the branch length calculated in Step 5c. In the example of Step 5b, that would be $(\overline{xz} + \overline{xy})/4 - \overline{yz}/2$.

Step 5 follows a simplified version of the method known as the unweighted pair group method with arithmetic mean (UPGMA). On how it applies to four sequences instead of three, see Lesk [82, pp. 164-165, Example 4.6]. UPGMA is useful as an illustration of distance-based tree estimation since it is a simplification of methods in common use that do not require the molecular clock hypothesis (Sect. 1.2.2). Specifically, the neighbor-joining method is described as a modification of UPGMA without that assumption [82, p. 165], and the neighbor-joining method in turn often closely approximates the minimum-evolution method [133, §3.3.3]. Both methods are considered forms of cluster analysis (Sect. 1.2.1).

3.1.2.2 Estimating a Tree from Protein Sequences

Each team that followed the steps of Sect. 3.1.1.1 shares the amino acid sequences from one of its trees with its opposing team (Step 1 of Sect. 3.1.1.2). The team receiving the amino acid sequences estimates the tree by following these ways to define a matrix of evolutionary distances:

(1) Use $n_{protein}$, which is the number of amino acid differences between each of the protein sequences, as the distance between those sequences in the distance matrix of Table 3.4.

- With that distance matrix, create the *protein-based uncorrected tree* by following the steps of Sect. 3.1.2.1.

(2) From each numeric value of $n_{protein}$ from Step 1 and with $N_{protein} = 4$ since there are 4 amino acids per protein sequence of Sect. 3.1.1, use $d_{protein}$, which is this *Poisson-corrected* [92, §2.2] distance between the sequences in the distance matrix of Table 3.4:

$$d_{protein} = N_{protein} \times \left| \ln\left(1 - \frac{n_{protein}}{N_{protein}}\right) \right|.$$

- The formula is intended to correct for the occurrences of multiple amino acid substitutions at the same site (amino acid position). A correction would be needed since the number of observed amino acid differences at a site cannot be more than 1 even if that site experienced 2 or more substitutions (Sect. 1.1).
- Note that the formula only works well when $n_{protein} / N_{protein}$ is small. For a very large numeric value of $n_{protein} / N_{protein}$, it can happen that $d_{protein} >$

$3 N_{protein}$, in which case $d_{protein}$ should be reset to $3 N_{protein}$ for using the method of Sect. 3.1.2.1 (cf. Step 2 of Sect. 3.1.2.3). The most extreme case of that is $n_{protein} / N_{protein} = 1$, which would yield $d_{protein} = N_{protein} |\ln 0| = \infty$.

- For real data, a large $n_{protein} / N_{protein}$ may indicate a lack of homology between the sequences or, as Hall [53, pp. 60–62] discusses, that they are poorly aligned with each other. Impressive phylogenetic trees can always be generated, but they only have evolutionary meanings when the sequences share a common ancestor, which implies that they are homologous (Sect. 1.2.1).
- While the proportion of substitutions cannot be greater than 1, allowing the higher maximum value of $d_{protein}$ works better for avoiding ties in the game.

• With that distance matrix, create the *protein-based corrected tree* by following the steps of Sect. 3.1.2.1.
• The formula does not correct for back substitutions (see Sect. 1.1), which are less of a problem for amino acid substitutions than for nucleotide substitutions.
• The formula is based on the model of the molecular clock as a Poisson process (see Sect. 2.1).

Notice how different the branch lengths are for the two protein-based trees even if they have the same shape. Two three-sequence trees have the same shape or *topology* if they agree on which of the three sequences diverged first from the other two.

3.1.2.3 Estimating a Tree from DNA Sequences

Next, each team shares the DNA sequence corresponding to the protein sequence of Sect. 3.1.2.2 with its opposing team (Step 2 of Sect. 3.1.1.2). The team receiving the DNA sequences estimates the tree by following these steps (based on Lesk [82, pp. 164-165, Example 4.6]) separately for each of these ways to define a distance or edge length:

(1) Use n_{DNA}, which is the number of nucleotide differences between each of the DNA sequences, as the distance between those sequences in the distance matrix of Table 3.4.

• With that distance matrix, create the *DNA-based uncorrected tree* by following the steps of Sect. 3.1.2.1.

(2) From each numeric value of n_{DNA} from Step 1 and with $N_{DNA} = 12$ since there are 12 nucleotides per DNA sequence of Sect. 3.1.1, use d_{DNA}, which is this

Jukes–Cantor-corrected [92, §3.2] distance between sequences in the distance matrix of Table 3.4:

$$d_{\text{DNA}} = N_{\text{DNA}} \times \left| \frac{3}{4} \times \ln \left(1 - \frac{4}{3} \times \frac{n_{\text{DNA}}}{N_{\text{DNA}}} \right) \right|. \tag{3.1}$$

- The formula is intended to correct for the occurrences of multiple nucleotide substitutions at the same site (nucleotide position). A correction would be needed since the number of nucleotide differences at a site cannot be more than 1.
- Note that the formula only works well when $n_{\text{DNA}} / N_{\text{DNA}}$ is small. For a very large numeric value of $n_{\text{DNA}} / N_{\text{DNA}}$, it can happen that $d_{\text{DNA}} > N_{\text{DNA}}$ or $n_{\text{DNA}} / N_{\text{DNA}} \geq 3/4$. In those cases, resetting d_{DNA} to N_{DNA} works well with the method of Sect. 3.1.2.1, for that creates ties that tend to avoid applying its Step 5 of Sect. 3.1.2.1 beyond the scope recommended by Hall [53, pp. 80–81], following Nei and Kumar [92]. For real data, a large $n_{\text{DNA}} / N_{\text{DNA}}$ may indicate a lack of homology between the sequences or, as Hall [53, pp. 60–62] discusses, that they are poorly aligned with each other.
- With that distance matrix, create the *DNA-based corrected tree* by following the steps of Sect. 3.1.2.1.
- The Jukes–Cantor correction applies not only to constant-rate models but also to those allowing the rate of substitution to vary over time (Sect. 3.5.3).

3.2 Relations of Different Types of Trees

The main ideas of Substituter are organized in Fig. 3.5 in terms of these six trees:

(1) Expected tree (Fig. 3.1)
(2) Realized tree (Sect. 3.1.1)
(3) Uncorrected protein-based tree (Sect. 3.1.2.1)
(4) Uncorrected DNA-based tree (Sect. 3.1.2.1)
(5) Corrected protein-based tree (Sect. 3.1.2.2)
(6) Corrected DNA-based tree (Sect. 3.1.2.3)

An *alignment* is a set of sequences edited to facilitate comparisons between their homologous building blocks. Each expected tree corresponds to one alignment: for each expected tree, all of the tip sequences together constitute one alignment.

Each alignment may be used with the steps of Sects. 3.1.2.1, 3.1.2.2, and 3.1.2.3 to estimate trees. Section 3.3 explains how to automate estimating trees from an alignment saved as a computer file. Uncertainty due to potential alignment errors is described in Sect. 3.4.1.

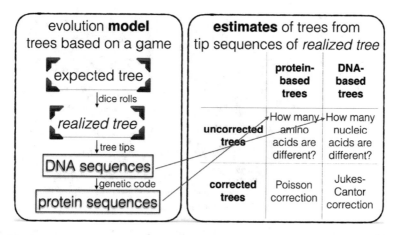

Fig. 3.5 Five trees from one expected tree

3.3   Software for Tree Estimation

These steps explain how to use MEGA X [74, 108] to estimate phylogenetic trees:

(1) Install MEGA X after downloading it from https://www.megasoftware.net.

(2) On your computer, find and double-click the "Crab_rRNA.meg" alignment file mentioned in the *MEGAX-Help* tutorial page at https://bit.ly/3lGqfgh ("Building Trees From Sequence Data").

(3) Examine the alignment to see which sequences are being compared and how they are positioned relative to each other.

(4) Follow the tutorial steps to construct a Neighbor-Joining (NJ) tree (Example 4.1 of the tutorial), for "Model/Method" selecting "p-distance." ("No. of differences" is what you used when you played Substituter, but if you divide it by the number of sites, you get the "p-distance.")

(5) Follow the same steps to construct a UPGMA tree instead of a NJ tree by clicking "PHYLOGENY" and then "Construct/Test UPGMA Tree. . ." instead of "Construct/Test Neighbor-Joining Tree. . ." (The UPGMA method is more like what you used when you played Substituter.)

(6) Select each window with a tree and click "View," "Show/Hide," and "Branch Lengths." That shows the numbers of differences.

(7) Follow the above steps except this time for "Model/Method" selecting "Jukes–Cantor model" instead of "p-distance." (That is the distance-correction method you used when you played Substituter when dividing by the number of sites.) This time, showing the branch lengths displays the corrected distances.

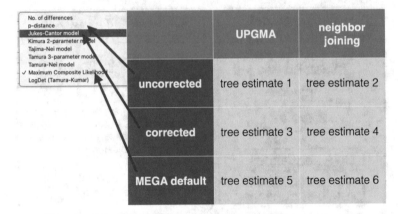

Fig. 3.6 2 distance methods × 3 substitution models = 6 estimates of trees. The first four tree estimates are generated by following the steps of Sect. 3.3. For more information on substitution models, see Sect. 3.5

By following those steps, you have created four of the trees mentioned in Fig. 3.6. To complete Exercise 3, you will also do so using the "mtCDNA.meg" and "Drosophila_Adh.meg" alignment files that came with MEGA X instead of the "Crab_rRNA.meg" alignment file. (In Substituter, your team generated a DNA alignment of three sequences for each of the expected trees (A, B, C, and D), for a total of four three-sequence alignments.)

3.4   Sources of Uncertainty in Tree Estimates

As seen in Chap. 1, each reconstructed tree is only an estimate based on many assumptions that are incorrect to varying and usually unknown degrees. Each source of uncertainty contributes to uncertainty about the estimated tree. The following sources of uncertainty have received the most attention.

3.4.1   Uncertainty About Common Ancestry

Phylogenetic tree estimation assumes the present-day sequences included in an alignment (Sect. 3.2) are homologous in the sense of having evolved from a common ancestor (Sect. 1.2.1). On the one hand, a high degree of sequence similarity, perhaps quantified by an extremely low E value in the BLAST software, suggests homology [53, chapter 3]. On the other hand, homology cannot be inferred from similarity alone, as clearly explained by Lesk [82, pp. 159–160].

Inadvertently including nonhomologous sequences in an alignment introduces errors in tree estimates that even affect the probabilities associated with the sequences in the alignment that are in fact homologous. Too much error of that type would mean the tree could not be interpreted in terms of evolutionary relationships (Sect. 1.2.1).

Xia [132, chapter 2] observes that many published phylogenetic trees suffer biases due to poor sequence alignment. Xia [132, chapter 2] further laments that journal editors tend to favor the publication of trees so large that peer reviewers cannot check all of the alignments. Even when care is taken to align the sequences as well as possible, different alignment algorithms make different assumptions, requiring researchers to try different algorithms on the same sequences in order to assess how much influence the algorithm has on the conclusions [132, chapter 2].

Any resulting uncertainty about the alignment increases the uncertainty about the estimated trees. That uncertainty is captured to some extent in Substituter since the sequences simulated from "Tree A" of Fig. 3.1 are incorrectly considered homologous by the team working in estimation mode (Sect. 3.1.2).

3.4.2   Uncertainty About the Topology

Uncertainty in the topology of a tree is often indicated by collapsing a bifurcation (e.g., Fig. 3.4) into a polytomy (e.g., Fig. 3.3). For example, that is what Step 4 of Sect. 3.1.2.1 does. Some of this uncertainty may be represented by the bootstrap measure of certainty to be covered in Sect. 4.1. It was not represented in Step 2 of Sect. 5.3 since the topology of the expected tree was used.

3.4.3   Uncertainty About the Branch Lengths

Even if the topology of part of a tree is correct, the branch lengths associated with that part cannot be known very precisely. Uncertainty about branch lengths leads to uncertainty about speciation rates and other biological quantities calculated from estimates of branch lengths. For example, uncertainty in the rate of evolution, even when neglecting other sources of uncertainty, would lead to errors in a measure of biodiversity from 10% to 38%, which is large enough to affect practical decisions about conservation [103].

Some of the uncertainty about branch lengths can be represented with confidence intervals, as will be seen in Sect. 4.2.2. It was not represented in Step 2 of Sect. 5.3 since the branch lengths of the expected tree were used. However, even confidence intervals corrected by the method of 4.1 fail to quantify all uncertainty since they depend on assumptions made by the substitution model.

3.4.4   Uncertainty About the Substitution Model

Every method of estimating a phylogenetic tree needs a mathematical model of substitutions. A simple model is the Poisson model that underlies the rules of Substituter (Sect. 3.1.1.1). Examples of more realistic models are mentioned in Sect. 3.5.

More realism in a model can improve estimation unless the model is too realistic in the sense of being too complex compared to the amount of data available. That is because accurately estimating numbers of substitutions even though many of them are hidden by other substitutions (Sect. 1.1) requires a substitution model that is neither too simple nor too complex. The model of substitution must be complex enough to capture the main features of molecular evolution and yet not overly complex, lest it fit the data so well that it does not generalize to past events [27].

Statistical tests are available to check the agreement of models with data. However, if all of the available substitution models fail the statistical tests, a researcher may react by discarding the sequences that lead to those test outcomes. Unfortunately, deleting those sequences may bias the results [27].

A portion of the uncertainty about the model can be assessed by trying different substitution models and noting how they affect the estimates [26, p. 448]. A more algorithmic solution is to compute the confidence interval of a branch length for each model and then to report the lowest and highest limits [cf. 19]. That approach is discussed further in Sect. 4.2.1. It may also be used to quantify the uncertainty about the tree estimation method.

A mathematical way to represent uncertainty about the model appears in Sect. A.3 of Appendix A.

3.4.5   Uncertainty About the Tree Estimation Method

Various methods of phylogenetic tree estimation include UPGMA, neighbor joining, and minimum evolution (Sect. 3.1.2.1). Some of this uncertainty can be assessed by trying different tree estimation methods and noting how they affect the topologies and branch lengths.

3.4.6   Uncertainty About the Statistical Method and Prior
Probabilities

Uncertainty about the statistical method and about prior probabilities (Sect. 1.2.3) tends to have large impacts on the uncertainty about the branch lengths and topologies of trees. These sources of uncertainty are described in Sects. 4.2.1 and 5.5.1.

Here, it is enough to note that while Bayesian posterior probabilities of clades have been reported to be misleadingly high [134], the bootstrap proportion (Sect. 4.1), often used with maximum likelihood methods, tends to be on the conservative side [19, 56, 136].

3.5 Excursus: Models of Nucleotide Substitution

The focus of this optional section is on nucleotide substitutions, but the statistical methods apply to amino acid substitutions with little modification [43, §13.1]. In practice, uncertainty about which substitution model to use propagates to uncertainty in the results (see Sect. 3.4.4).

3.5.1 Background Terminology

Recall from Sect. 1.1 that a *substitution* is a change of which amino acid or nucleotide at a certain protein or DNA site is predominant in a given population. A *transition* is the substitution of one purine (adenine or guanine) for the other purine or of one pyrimidine (cytosine or thymine) for the other, whereas a *transversion* is the substitution of a purine for a pyrimidine or vice versa.

3.5.2 Discrete-Time Models [43, §13.2]

These models treat time as discrete in the sense of being represented in integer numbers of time units:

- *Discrete-time Jukes–Cantor model.* This model has only one parameter, the nucleotide substitution rate. The *simple symmetric PAM model* is the amino acid version of the Jukes–Cantor model. Other PAM models of amino acid substitutions are more realistic.
- *Kimura models.* The original Kimura model has two parameters, the rate of transitions and the rate of transversions, the latter of which is lower. There is also a three-parameter Kimura model. Another three-parameter generalization has different rates for purine-to-pyrimidine and pyrimidine-to-purine substitutions. Generalizations with more than three parameters have also been suggested.
- *Felsenstein model.* Felsenstein proposed a model that generalizes both the Jukes–Cantor model and a Markov chain. The Felsenstein model is *reversible* in the sense that it has the same stationary distribution backward in time as it does forward in time [43, §10.2.4]. Reversibility is important when comparing two sequences with an unknown common ancestor since in that case the direction of

time cannot be determined for all substitutions. A reversible generalization of the
Felsenstein model has been proposed.
- *HKY model.* The HKY model generalizes both the two-parameter Kimura model
  and the Felsenstein model.
- *Rate-varying models.* More complex models do not assume that all sites on a
  sequence evolve at the same rate. A popular model of this type assumes a gamma
  distribution of rates.

3.5.3 Continuous-Time Models [43, §13.3]

These models drop the assumption of discrete time:

- *Continuous-time Jukes–Cantor model.* Numbers of substitutions in the
  continuous-time Jukes–Cantor model [64] follow the homogeneous (constant-
  rate) Poisson distribution. This model corrects for multiple and back substitutions
  (see Sect. 1.1) in the estimation of the number of substitutions that took place
  since divergence from a common ancestor, as follows. Given the proportion of
  sites that differ between two sequences, one can use Eq. (3.1) to estimate the
  number of substitutions.

  - Bickel and West [20] used a doubly stochastic Poisson process to demonstrate
    that Eq. (3.1) also applies to the case of a rate of evolution that fluctuates in
    time, assuming that the mean rate of substitution is both stationary and the
    same for transitions and transversions.
  - Nei and Kumar [92, p. 37] derived equation (3.1) from the discrete-time
    Jukes–Cantor model mentioned in Sect. 3.5.2.

- *Other continuous-time models.* There are continuous-time versions of the
  discrete-time models described in Sect. 3.5.2.

3.5.4 Further Reading

Section 3.5 closely follows Ewens and Grant [43, chapter 13], which may be
consulted for details and additional references to the primary literature. That book
is recommended in Sect. 7.1.2.

3.6 Exercises

(1) After identifying the members of two competing teams, work with your team
    to defeat the opposing team.

(a) After following the rules of Sect. 3.1.1, your team will work in simulation mode to generate the sequence data the opposing team needs for Exercise 1b.

- It is strongly recommended that all sequences and trees generated in the following steps are clearly organized in a cloud service such as Google Docs or another location that is easily accessible by all members of your team. Depending on what technology you use for that, clear organization may involve detailed file or note names, folders or directories, and tags or labels.

  - That will help not only with the current level of Substituter but also with the next level. The rules for the next level are given in Sect. 5.2.
  - This recommendation also applies to scanned copies of any drawings made on paper.

(b) Follow the rules of Sect. 3.1.2 in estimation mode for an alignment provided by the opposing team. Your team will then be ready for these exercises:

  (i) Organize the four **estimated trees** you generated as follows. Draw two lines to divide a blank sheet of real or virtual paper into four quadrants, like those of the right-hand side of Fig. 3.5:

    (A) In the top-left quadrant, draw your uncorrected protein-sequence tree. Fill in its uncorrected branch lengths and tip protein sequences. Next to the tree, put its uncorrected distance matrix.
    (B) In the bottom-left quadrant, draw your corrected protein-sequence tree. Fill in its corrected branch lengths and tip protein sequences. Next to the tree, put its corrected distance matrix.
    (C) In the top-right quadrant, draw your uncorrected DNA-sequence tree. Fill in its uncorrected branch lengths and tip DNA sequences. Next to the tree, put its uncorrected distance matrix.
    (D) In the bottom-right quadrant, draw your corrected DNA-sequence tree. Fill in its corrected branch lengths and tip DNA sequences. Next to the tree, put its corrected distance matrix.

  (ii) Are the topologies of your two protein-based trees (one uncorrected and the other corrected) the same or different?
  (iii) Are the topologies of your two DNA-based trees (one uncorrected and the other corrected) the same or different?
  (iv) Are the topologies of your two corrected trees (one protein-based and the other DNA-based) the same or different?

(v) Which of the four expected trees of Fig. 3.1 do you think the opposing team used to generate the sequences it gave you? Why?

(vi) Organize the **endgame** as follows. Draw two lines to divide a blank sheet of real or virtual paper into four quadrants.

   (A) In the top-left quadrant, copy your DNA-based corrected tree (with its branch lengths and tip DNA sequences) from Step 1(b)iD.

   (B) In the bottom-left quadrant, draw what you guess is the expected tree, knowing it is one of the trees displayed in Fig. 3.1.

   (C) Tell the opposing team which expected tree you just guessed in Exercise 1(b)viB. After that, ask the opponent which expected tree is correct, and then draw it in the top-right quadrant.

   (D) In the bottom-right quadrant, draw the realized tree provided by the opponent (with its branch lengths and with the DNA sequences of its tips).

(c) Who won? Scoring for the DNA-based corrected trees:

   (i) **Gain** 10,000 points for each correctly guessed expected tree.

   (ii) Section 3.2 and Fig. 3.5 explain what the "realized tree" is. Does your DNA-based corrected tree have the same topology as the realized tree?

      (A) **If so**, then **lose** 1000 points for every substitution that the branch lengths of your expected DNA-based tree differ from those of the realized tree.

      (B) **If not**, then **lose** 10,000 points.

(d) **Estimated trees versus the true tree:**

   (i) What does playing Substituter tell you about the distinction between the realized tree and an estimated tree? Again, Sect. 3.2 and Fig. 3.5 explain what the "realized tree" is.

   (ii) Which sources of uncertainty listed in Sect. 3.4 contributed to the differences you saw in Exercise 1(c)ii between the realized tree and the estimated tree?

   (iii) Which sources of uncertainty listed in Sect. 3.4 are in nature but not in Substituter?

(e) **More takeaways from Substituter:**

   (i) What is the purpose of each of the six trees listed in Sect. 3.2? Hint: Fig. 3.5.

   (ii) Which tree corresponds to the process of molecular evolution?

   (iii) Which four trees are only estimates of that tree?

Fig. 3.7 Number of amino
acids different = number of
amino acid sites in the
alignment − number of
amino acids the same. Data
source: Lesk [82, p. 158,
Case Study 4.4]

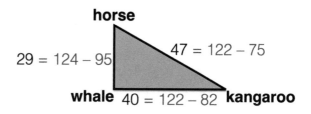

(iv) How does the remaining tree relate to the molecular clock hypothesis (Sect. 1.2.2)?

(2) **Real data:**

(a) Create both the protein-based uncorrected tree and the protein-based corrected tree for the 3 sequences of Fig. 3.7.

(b) Are their topologies the same or different?

(c) What are the branch lengths?

(d) How close do you think your estimated trees are to what really happened in evolution? Hint: review the sources of uncertainty listed in Sect. 3.4 and your answer to Exercise 1d.

(3) Follow the steps of Sect. 3.3 to answer these questions **for each of the alignment files** it mentions:

(a) How does applying the Jukes–Cantor correction affect the neighbor-joining tree?

(i) How does it affect the topology (which sequences are in the same cluster)?

(ii) How does it affect the branch lengths?

(b) How does applying the Jukes–Cantor correction affect the UPGMA tree?

(i) How does it affect the topology?

(ii) How does it affect the branch lengths?

(c) Considering only the trees using the Jukes–Cantor correction, compare the neighbor-joining tree to the UPGMA tree.

(i) Do they have the same topology?

(ii) How different are their branch lengths?

(d) This exercise illustrates some of the sources of uncertainty listed in Sect. 3.4.

(i) Which of those sources of uncertainty are considered?

(ii) Which of those sources of uncertainty are not considered?

Chapter 4
Estimating Divergence Times

> *In the space of one hundred and seventy-six years the Lower Mississippi has shortened itself two hundred and forty-two miles. That is an average of a trifle over one mile and a third per year. Therefore, any calm person, who is not blind or idiotic, can see that in the Old Oolitic Silurian Period, just a million years ago next November, the Lower Mississippi River was upwards of one million three hundred thousand miles long, and stuck out over the Gulf of Mexico like a fishing-rod. And by the same token any person can see that seven hundred and forty-two years from now the Lower Mississippi will be only a mile and three-quarters long, and Cairo and New Orleans will have joined their streets together, and be plodding comfortably along under a single mayor and a mutual board of aldermen. There is something fascinating about science. One gets such wholesale returns of conjecture out of such a trifling investment of fact.*
>
> – Mark Twain[1]

A *divergence time* is the amount of time that elapsed since the ancestors of present-day sequences began different paths of evolution from their common ancestor. It tells us how long ago the most recent ancestor of the sequences would have existed.

Divergence times are typically estimated using the branch lengths of an estimated phylogenetic tree like one of Chap. 3. The fossil record is often used for calibration, transforming branch lengths to divergence times. A *time tree* is a phylogenetic tree with divergence time estimates [53, chapter 15].

The concepts of divergence times and time trees were explained in Sect. 1.1 using an example featuring three variants of a virus. This chapter addresses the need to

---

Electronic Supplementary Material The online version contains supplementary material available at (https://doi.org/10.1007/978-3-031-11958-3_4).

---

[1] *Life on the Mississippi* [119].

D. R. Bickel, *Phylogenetic Trees and Molecular Evolution*, SpringerBriefs in Systems Biology, https://doi.org/10.1007/978-3-031-11958-3_4

propagate uncertainty in the model and clade to phylogenetic trees constructed from sequence data.

4.1 Confidence Intervals of Divergence Times

Many computer programs for computing divergence times report some of the uncertainty about them in terms of *confidence intervals* in addition to single numbers called *point estimates*. A confidence interval depends on a level, usually 95%. For example, instead of reporting a divergence time as a point estimate such as 82 MY (82 million years ago), we might report [28 MY, 230 MY] as the 95% confidence interval [19]. The 28 MY and the 230 MY are called the *lower limit* and the *upper limit* of the confidence interval.

If all of the assumptions in the estimation method were correct, the 95% level is the proportion of simulated data sets for which the true divergence time would be between the limits of the confidence interval computed on the basis of each data set. That could only be interpreted as an exact probability that the true divergence time is in the interval if all of the assumptions in the estimation method were correct, which is never the case. The uncertainty about many of the assumptions can be quantified by lowering the level of the interval, as will be seen in Sect. 4.2.

For now, we will only reduce the confidence level to address the uncertainty about the clade that gives the divergence time its meaning. A level of certainty of each clade can be estimated by the *bootstrap proportion* (92, chapter 9; 53, chapter 6), which is available as an option in MEGA [74, 108]. The bootstrap proportion is a number between 0 and 1 that tends to be lower when there is more uncertainty about whether the clade of the estimated tree is correct. Since the bootstrap proportion estimates the probability that the clade is correct, it may be multiplied by the confidence level in order to adjust it for uncertainty about the clade [19]. For example, if the [28 MY, 230 MY] confidence interval depends on a clade with a bootstrap proportion equal to 84%, then, since $0.84 \times 0.95 = 0.80$, we would report [28 MY, 230 MY] as an 80% confidence interval rather than as a 95% confidence interval [19]. For additional examples or for guidance for using software to correct not only the level but also the interval itself, try Exercise 1. The reasoning behind the multiplication rule is explained in Sect. 4.3.

4.2 Uncertainty in Divergence Time Estimates

4.2.1 Sources of Uncertainty in Divergence Time Estimates

Uncertainty in estimates of divergence time (Sect. 4.1) calls for "healthy skepticism" regardless of how big the data set is [27, Boxes 1, 3]. Here are some of the many sources of uncertainty:

4.2.1.1 Uncertainty About Phylogenetic Trees

The general sources of uncertainty listed in Sect. 3.4 introduce uncertainty in the topologies and branch lengths needed to estimate divergence times. For example, the assumption of a common ancestor (Sect. 1.2.1) is crucial to correctly aligning the sequences. Evolutionary conclusions drawn from incorrect alignments are invalid (Sect. 3.4.1), and yet errors in automated alignments cannot be checked manually if the number of sequences is too large [27].

Even given a correct alignment, there is also considerable uncertainty in trees due to uncertainty in substitution models (Sect. 3.4.4). Methods of estimating divergence times that do not rely on the molecular clock hypothesis instead rely heavily on model assumptions [69]. Estimating divergence times from trees constructed from molecular sequences relies on model assumptions mostly made to enable calculations instead of being based on actual observations [103]. For example, uncertainty about the substitution model can lead to differences of divergence time estimates as large as hundreds of millions of years in the cases of animals and by almost a billion years in the case of cyanobacteria [27].

Uncertainty about the model is not reflected in confidence intervals since each confidence interval depends on the assumptions behind the model used to compute it [27]. Bromham et al. [28] recommend reporting the range of results from using different models. Confidence intervals from different models can then be combined by reporting a single interval with its lower limit as the minimum of the models' lower limits and its upper limit as the maximum of the models' upper limits (Exercise 6a). That agrees with the method of Bickel [19] whenever the confidence intervals overlap but is more cautious in other cases. Exercise 6b illustrates the less conservative method of Bickel [15].

The confidence intervals may incorporate some of the uncertainty about trees when the level of confidence is corrected using the method of Sect. 4.1. For divergence time estimation, Yang [133, §10.3.1] recommends methods that make the guesses needed to construct bifurcating trees (e.g., Fig. 3.4) rather than methods that represent uncertainty in the topology by polytomies (e.g., Fig. 3.3).

The Bayesian analog of a confidence interval is called a *credible interval*. Fully Bayesian models would be ideal for propagating uncertainty were their assumptions true. However those assumptions include *prior distributions*, which are probability distributions that are not estimated from data but rather are given as input to Bayes's theorem (Sect. 1.2.3).[2] Uncertainty resulting from the use of prior distributions is discussed in Sect. 4.2.1.5. Narrow credible intervals do not necessarily indicate less uncertainty, for they can instead indicate conflicts between the model and the data [41]. Another indication of conflict with the data is too much agreement between the prior distributions and the *posterior distributions*, the data-dependent probability distributions that update the prior distributions [27]; see Sect. 4.2.1.5.

---

[2] Exception: estimating prior distributions from data is the defining characteristic of empirical Bayes methods (Sect. 7.2).

4.2.1.2   Uncertainty About Fluctuations in the Substitution Rate

Variations in the rate of evolution contribute to uncertainty about divergence times [103]. The simplest methods of estimating divergence times must assume the molecular clock hypothesis that was discussed in Sects. 1.2.2 and 3.1.2.1. More complicated methods allow the rate of substitution to change in various ways but then must rely on assumptions about how the rate might have changed over time. Since we do not know how the substitution rate has changed over time, it must be modeled mathematically, as, for example, for Fig. 1.1. Different models account for different sources of uncertainty about fluctuations in the substitution rate (Sect. 1.2.2; Yang 133, §10.3.3). For example, uncertainty about whether the scale-free models of Sect. 2.8 accurately describe molecular evolution increases the uncertainty about results from scale-dependent models.

Substitution rates can vary substantially even for very similar species, and variations in those rates cannot necessarily be accurately estimated [27]. Since, given enough sequences, the substitution rate is approximately the number of substitutions divided by the time interval between two nodes on the tree, it could in theory be estimated given a reliable estimate of the number of substitutions and a reliable estimate of the time interval based on the fossil record. When the time interval is unknown, it can be accurately estimated only given both a reliable estimate of the number of substitutions and a reliable estimate of the rate. That rate in turn could only be an estimate with some probabilistic assumptions about how the rate might differ from the rate estimated on the parts of the tree covered by the fossil record. Those assumptions underly methods based on relaxed clock models [27]. As noted in Sect. 3.4.4, more complex models do not necessarily lead to better results. Those models can be tested to some extent, but passing a test is not evidence that the model is reliable [27].

While more sequence data can reduce uncertainty in branch lengths, that does not reduce uncertainty in divergence times due to variations in rates of evolution between branches [114]. In fact, divergence times are more sensitive to changes in model assumptions and in fossil calibrations than to increasing amounts of molecular data [79]. More generally, since divergence times are not observed but rather are historical inferences, increasing the number of sequences used in the data analysis does not necessarily reduce the uncertainty [27].

Gillespie [47] emphasized that even models of molecular evolution that do not assume the clock rely on the assumption that the rate of evolution is stationary in the sense that that the statistical properties of rate fluctuations do not change in time. For example, all of the models represented in Fig. 1.1 imply stationarity. Because sequences are usually only available for present-day organisms, stationarity cannot be tested. Fortunately, stationarity need not be assumed for the entire history of a sequence, but only for the period of time studied.

Just as substitution rates vary across a phylogenetic tree, so do the rates of the birth and death of species. Many analyses overlook that, drawing conclusions about variations in speciation rates without testing the corresponding null hypothesis that the rates do not really change [27]. Conclusions about macroevolution can be misled by variations in speciation rates that were not included in the models that led to those conclusions [27].

4.2.1.3 Uncertainty About the Dates and Relations of Fossils

Since estimated dates of fossils are needed for calibration, uncertainty in those dates propagates to estimates of divergence times [133, §10.3.4.1]. While maximum likelihood estimation (Sect. 5.4) ignores this source of uncertainty, more sophisticated methods at least address it [133, §10.3.4.2]. In some cases, most of the uncertainty about divergence times comes from uncertainty about estimated dates of events in the fossil record, and there is always considerable uncertainty about those dates since they are necessarily based on assumptions about the distant past rather than on experiments that can be controlled in the laboratory [27].

Not only are the dates of events in the fossil record uncertain, but so are their positions on the tree [103]. Due to all the uncertainties involved, caution is needed when interpreting divergence time estimates from both the fossil record and sequence data [26, p. 459]; see Sect. 2.6.

4.2.1.4 Uncertainty in Statistical Error

Neglecting confidence intervals is a source of uncertainty that can lead to serious errors in divergence time estimates [40, p. 3]. For more on confidence intervals, see Sect. 4.2.2.

4.2.1.5 Uncertainty About Prior Distributions

As mentioned in Sect. 4.2.1.1, the Bayesian alternative to the confidence interval is the credible interval. Bayesian methods of estimating divergence times have the advantage of propagating multiple sources of uncertainty to the credible intervals.

That advantage, however, requires the specification of prior distributions that are themselves uncertain, and that source of uncertainty is not reflected in the credible intervals [28]. For example, changing prior distributions doubles the estimates of the divergence times of placental mammals [27]. Unavoidable biases in how sequences are sampled affect Bayesian methods to an extent depending on how their assumed prior distributions model the sampling procedure [27]. When the posterior distribution of divergence times agrees with the prior distribution, that may indicate that the sequence data do not affect the conclusions enough for the analysis to be considered reliable [27].

Uncertainty in the conclusions resulting from uncertainty about the prior distribution can be assessed by noting how changing that assumptions about the prior distribution affects the conclusions [27]. Since changing the prior distribution can lead to completely different conclusions, a non-Bayesian method may be used to assist in selecting a prior distribution and may in some settings be used as an alternative to a Bayesian method [7].

The confidence intervals from non-Bayesian methods and the credible intervals from Bayesian methods may then be combined by the method of taking the extremes of their limits (Sect. 4.2.1.1) [19] or by a less conservative method [15]. Those methods power the software of Exercises 1 and 6 (Sect. 4.4).

4.2.2 Quantified Uncertainty in Divergence Time Estimates

Recall that *clades* are clusters assumed to have a common ancestor in a phylogenetic tree (Step 5b of Sect. 3.1.2.1). The *divergence time estimate* between clades is a guess at how many years ago they might have separated from their most recent common ancestor. When multiple estimates of relevant fossil dates are available, it is possible to report a 95% confidence interval (Sect. 4.1) rather than a single number as the divergence time estimate.

Example 1 If the fossil record suggests $t_{younger}$ and t_{older} as two divergence time estimates that satisfy $t_{younger} < t_{older}$, then an approximate 95% confidence interval is

$$\frac{t_{younger} + t_{older}}{2} \pm 6 \times \left(t_{older} - t_{younger}\right)$$

according to the assumption that they are independent and normally distributed. ▲

Confidence intervals address the uncertainty in statistical error, which is one of the sources of uncertainty listed in Sect. 4.2.1. Once a 95% confidence interval has been computed, its nominal confidence level of 95% can be corrected for other sources of uncertainty, as follows.

4.2.3 Correcting Unquantified Uncertainty in Divergence Time Estimates

Having stressed some of the sources of uncertainty mentioned above (Sect. 4.2.1), Donoghue and Smith [40, p. 3] warned "Unless all of these sources of error are taken into account, in addition to attempts to correlate fossil occurrences to the geological timescale, and those errors attendant to molecular clocks themselves, errors will propagate, potentially beyond the age of the events being estimated." Many of those issues remain unresolved by current models, leaving unanswered questions about the evolution of animals [26, pp. 457–460] that were raised in the 1990s (Sect. 2.4), in spite of Aris-Brosou and Yang [5] and later studies.

One way to address that problem is to correct the 95% confidence level of confidence intervals (Sect. 4.2.2) using the concept of the *proportion of unquan-*

tified uncertainty [cf. 13, 14]. That proportion of uncertainty not captured by the confidence interval is a percentage abbreviated by the letter u. For example, if the 95% confidence interval fails to account for half of the relevant uncertainty, then $u = 50\%$.

More precisely, u is the probability that certain results of data analysis do not apply. Said differently, $100\% - u$, the *proportion of quantified uncertainty*, is the probability that the results do apply [19]. That probability is multiplied by probabilities in the results to correct them for unquantified uncertainty. For example, we saw in Sect. 4.1 that the bootstrap proportion is multiplied by the confidence level to adjust it for some of the unquantified uncertainty about the clade.

The *corrected confidence level* is reduced from 95% to

$$(100\% - u) \times 95\% = (1 - u) \times 0.95.$$

In the case of the $u = 50\%$ example, the corrected confidence level is

$$(100\% - 50\%) \times 95\% = (1 - 0.5) \times 0.95 = 0.48 = 48\%.$$

That means that we would be only 48% sure that the divergence time is in the calculated confidence interval. That is much less than the 95% confidence interval we would have if there were no unquantified uncertainty ($u = 0\%$).

4.3   Excursus: Why Multiply Probabilities by the Proportion of Quantified Certainty?

The proportion of quantified uncertainty (Sect. 4.2.3) is the probability that the model assumptions are close enough to the truth for practical purposes. Why should that probability be multiplied by the probabilities that are based on the assumptions?

Uncertain models report a result and the conditional probability that the result holds given the condition that the assumptions of the model are true or at least adequate for practical purposes. A conservative estimate of the probability of the result is the *joint probability* both that the result holds and that the assumptions are an adequate approximation. That joint probability is the conditional probability multiplied by the probability of adequate assumptions, according to the definition of conditional probability (see Sect. 5.1). Putting it all together, a conservative estimate of the probability of the result is the conditional probability multiplied by the proportion of quantified uncertainty.

That explains why the 95% confidence level (Sect. 4.1) and other probabilities of results (Sects. 4.2.3 and 5.5.2) are multiplied by the bootstrap proportion or by another estimate of $100\% - u$. Bickel [19] provides both Bayesian and non-Bayesian justifications of this method of propagating uncertainty.

Example 2 If the only sources of uncertainty were the uncertainty of the divergence time given the topology, the uncertainty of the topology given the homology of the aligned sequences, and the uncertainty of the homology, then the corrected confidence level of the 95% confidence interval of the divergence time would be the product of the conditional probabilities of those three assumptions. Mathematically, using the "|" character to abbreviate "conditional on":

$$\text{Pr (time is in 95\% interval)} = \text{Pr (time is in 95\% interval}|\text{topology)} \times \text{Pr (topology)}$$

$$= \text{Pr (time is in 95\% interval}|\text{topology)} \times \text{Pr (topology}|\text{homology)} \times \text{Pr (homology)}$$

$$= 95\% \times \text{Pr (topology}|\text{homology)} \times \text{Pr (homology)}$$

$$= 95\% \times (100\% - u)$$

$$= 0.95 \times (1 - u),$$

where $u = 100\% - \text{Pr (topology}|\text{homology)} \times \text{Pr (homology)}$. However, that is only a lower bound on u since those are not the only sources of uncertainty. As a result, the corrected confidence level computed in that way is only an upper bound on the probability that the divergence time is in the 95% confidence interval. ▲

For technical details on this kind of correction of unquantified uncertainty, see Sect. 7.3, which mentions a sense in which the corrected probability is a *lower* bound. The nature of the uncertainty-corrected probability is further clarified in Appendix A.

4.4   Exercises

(1) The following questions ask about the tree estimation result that is presented in Table 4.1.

   (a) For each of the four clades, what is the uncertainty-corrected level of the confidence interval for the divergence time after adjusting for uncertainty about the clade? Hint: for clade A, the corrected level is worked out in Sect. 4.1.

Table 4.1 Divergence time 95% confidence intervals and bootstrap proportions for a phylogenetic tree of bacterial species [19] as estimated by the 3-parameter model of Tamura [113]

Clade in the estimated tree	Divergence time (95% confidence interval)	Bootstrap proportion
A	[28 MY, 230 MY]	84%
B	[31 MY, 250 MY]	32%
C	[37 MY, 250 MY]	81%
D	[66 MY, 370 MY]	90%

(b) For each of the four clades, determine the 68% uncertainty-corrected confidence interval of the divergence time by following these steps:

   (i) Open https://davidbickel.shinyapps.io/NormalUncertainty/ [19] in a web browser.

   (ii) Enter the lower and upper limits of the 95% confidence interval given in Table 4.1.

   (iii) Enter the bootstrap proportion given in Table 4.1 as a conservative estimate of the probability that the clade exists.

   (iv) Select the button corresponding to the goal of obtaining an interval that is 68% sure of containing the divergence time, and enter that probability into the corresponding box.

   (v) Read the result. What to do next depends on whether the result is an "Error" or an interval:

      (A) If an error message indicates that 68% is too high for that clade, explain why its estimated probability of existing is incompatible with having 68% certainty in its divergence time.

      (B) If there is no error message, then copy the resulting 68% uncertainty-corrected confidence interval of the divergence time.

(c) The *reference clade* is the clade used in this step to set the uncertainty-corrected confidence level for other clades. Among the three clades that have 68% uncertainty-corrected confidence intervals according to what you found in Exercise 1b, which clade has the lowest of the three bootstrap proportions? Mark it as the reference clade. Follow these steps for each of those three clades, including the reference clade:

   (i) Reload https://davidbickel.shinyapps.io/NormalUncertainty/ [19].

   (ii) Enter the lower and upper limits of the 95% confidence interval given in Table 4.1.

   (iii) Enter the bootstrap proportion given in Table 4.1 as a conservative estimate of the probability that the current clade exists.

   (iv) Select the option button corresponding to the use of a reference clade.

   (v) Enter the bootstrap proportion given in Table 4.1 as a conservative estimate of the probability that the *reference clade* exists. (This will be the same value as that of Step 1(c)iii when the current clade is the reference clade.)

   (vi) Copy the result.

(d) In Exercise 1c, what is an advantage of considering the clade with the lowest of the three bootstrap proportions as the reference clade? Hint: try using other clades as the reference clade.

(e) What are the advantages and disadvantages of each of these ways you incorporated uncertainty about the clade?

(i) Adjusting the confidence level for unquantified uncertainty (Exercise 1a)
(ii) Determining the 68% uncertainty-corrected confidence interval (Exercise 1b)
(iii) Using a reference clade (Exercise 1c)

(2) Why should a correction for sources of uncertainty about a divergence time that are not represented in a 95% confidence interval (Sect. 4.2.1) always result in a confidence level that is *lower* than 95% (Sect. 5.5.2)? In other words, why should not the uncertainty correction ever make the confidence level *higher* than 95%?

(3) These questions show two ways to use the "fungi" time tree of http://timetree. org [75] to generate a 95% confidence interval for divergence time estimation (Sect. 4.2.2):

(a) Find a node on the tree that explicitly reports a confidence interval ("CI"). What is that confidence interval?
(b) Find a node on the tree that instead reports minimum and maximum divergence times ("RANGE"). Interpreting those times as $t_{younger}$ and t_{older}, what is the corresponding confidence interval according to Example 1 of Sect. 4.2.2?

(4) According to Sect. 4.2.3's correction of the 95% level of the confidence intervals of Exercise 3:

(a) What is the value of the corrected confidence level using $u = 25\%$? Hint: these questions are answered for $u = 50\%$ in Sect. 4.2.3.
(b) What is the value of the corrected confidence level using $u = 75\%$?
(c) What is the value of the corrected confidence level using $u = 100\%$? Is that value of the corrected confidence level what you would expect if the 95% confidence interval failed to account for *any* of the relevant uncertainty?
(d) What is the value of the corrected confidence level using $u = 0\%$? Is that value of the corrected confidence level what you would expect if the 95% confidence interval successfully accounted for *all* of the relevant uncertainty?

(5) These foundational questions ask you to give more thought to the values of unquantified uncertainty mentioned in Exercise 4 ($u = 0\%, 25\%, 50\%, 75\%, 100\%$):

(a) In your opinion, which of those values of u would be most appropriate for divergence times measured in numbers of months, as in viral evolution? What about for divergence times in hundreds of MYs, using the fossil record? How would you defend your answers? Hint: keep in mind the definition of the proportion of unquantified uncertainty (u) given in Sect. 4.2.3 while carefully weighing the sources of uncertainty mentioned in Sect. 4.2.1, including those listed in Sect. 3.4. Considering the bootstrap proportion of Sect. 4.1 as an upper bound on $100\% - u$ may also improve

Table 4.2 95% confidence intervals of the divergence times for the bacterial clades estimated as described by Bickel [19]. Whereas "Model 1" is the same model as that of Table 4.1, "Model 2" is the general time-reversible model of Nei and Kumar [92]

Clade in the estimated tree	Model 1	Model 2
A	[28 MY, 230 MY]	[36 MY, 270 MY]
B	[31 MY, 250 MY]	[44 MY, 320 MY]
C	[37 MY, 250 MY]	[55 MY, 330 MY]
D	[66 MY, 370 MY]	[93 MY, 460 MY]

your reasoning about this. That upper bound may need to be multiplied by other probabilities, as seen in Example 2.

(b) What sources of uncertainty are not mentioned in Sect. 4.2.1? Hint: review the history outlined in Chap. 2 and its epigraph. How much would those additional sources increase the value of u that you estimated in Exercise 5a?

(6) The following questions ask about the tree estimation result that is presented in Table 4.2.

(a) Uncertainty about whether to report the confidence interval under the "Model 1" column or the "Model 2" column is an example of uncertainty about the substitution model (Sect. 3.4.4). For each of the four estimated clades, what is the smallest interval that includes all the divergence times in the intervals for both models? Hint: that is the method of uncertainty propagation explained under Sect. 4.2.1.1.

(b) For each of the four clades, follow this method of determining the uncertainty-corrected credible interval:

   (i) Open https://davidbickel.shinyapps.io/MixtureUncertainty/ [15] in a web browser.

   (ii) In the box for the **lower** limits, enter these in order:

      (A) The number representing the lower limit for Model 1
      (B) A comma (",")
      (C) The number representing the lower limit for Model 2

   (iii) In the box for the **upper** limits, enter these in order:

      (A) The number representing the upper limit for Model 1
      (B) A comma (",")
      (C) The number representing the upper limit for Model 2

   (iv) Copy the resulting uncertainty-corrected credible interval.

(c) For each of the four clades, does Exercise 6a method or Exercise 6b method give wider credible intervals? In your opinion, which method more reliably propagates uncertainty about the model? Defend your answer. Hint: Sect. 7.3 may have some ammunition.

Chapter 5
Estimating Common Ancestors

Notions of significance tests, confidence intervals, posterior intervals and all the formal apparatus of inference are valuable tools to be used as guides, but not in a mechanical way; they indicate the uncertainty that would apply under somewhat idealized, may be [sic] very idealized, conditions and as such are often lower bounds to real uncertainty.

– Sir David Cox[1]

The problem of guessing hypothetical ancestral sequences is described using game rules in Sect. 5.2.1. Two methods of estimation are then described in Sects. 5.3 and 5.4.

5.1 Notation for Conditional Probability

This section reviews conditional probability, a concept used in Sect. 4.3, and also introduces some symbols that will be used in equations to come.

Suppose you have 1 fair 6-sided die. Then the probability that you will roll a 5 is $1/6$, about 17% or 0.17. A convenient way to abbreviate that sentence is

$$\Pr(1d6 = 5) = \frac{1}{6} \approx 17\% = 0.17. \tag{5.1}$$

Now suppose you keep rolling the die until you get one of the three odd numbers. Then the *conditional probability* that you will roll a 5, given the condition that you

---

Electronic Supplementary Material The online version contains supplementary material available at (https://doi.org/10.1007/978-3-031-11958-3_5).

[1] *Statistical Science* [35]. Reprinted with the permission of the Institute of Mathematical Statistics.

roll an odd number, is $1/3$, about 33% or 0.33. That sentence is abbreviated by

$$\Pr(1d6 = 5|1d6 = 1 \text{ or } 3 \text{ or } 5) = \frac{1}{3} \approx 33\% = 0.33. \tag{5.2}$$

The game changes, and you are to keep rolling the die until you roll 5 or 6. Then the conditional probability that you will roll a 5, given the condition that you roll 5 or 6, is $1/2$, exactly 50% or 0.5. That sentence is abbreviated by

$$\Pr(1d6 = 5|1d6 = 5 \text{ or } 6) = \frac{1}{2} = 50\% = 0.5. \tag{5.3}$$

Why do you need the condition after the "|" character, called the *vertical bar*? Because without it, Eqs. (5.2) and (5.3) would say $\Pr(1d6 = 5) = 1/3$ and $\Pr(1d6 = 5) = 1/2$, which would mean $1/3 = 1/2$. Contradictions like that and other errors can be avoided by clearly stating the condition after the vertical bar whenever we write a conditional probability. On the other hand, in complex problems, it is not practical to keep repeating all the conditions assumed in the data analysis.

Since this chapter is about inferring ancestral traits, suppose that you know three girls who have the same grandmother and that only two of them are sisters. The first sister has blue eyes, but the second sister and the cousin have brown eyes. What is the conditional probability that the grandmother has blue eyes given the conditions that the first sister has blue eyes but that second sister and the cousin have brown eyes? That conditional probability could be designated by

Pr (ancestor's eye color = blue|first sister's eye color

= blue, second sister's eye color

= brown, cousin's eye color = brown) ,

but that is too long to be useful. Thinking of the girls as tips in the tree of Fig. 3.4 and dropping "eye color," we could instead abbreviate the conditional probability by

$$\Pr\left(\text{ancestor} = \text{blue}|\text{tip}_1 = \text{blue}, \text{tip}_2 = \text{brown}, \text{tip}_3 = \text{brown}\right).$$

That is more like the generic notation used in the next section.

5.2   Estimating Ancestors in Substituter

5.2.1   Estimating an Ancestral DNA Sequence

This section gives the rules of the next level of Substituter, the dice game introduced in Sect. 3.1. Based on an alignment of three DNA sequences provided by the opposing team, your team will guess the DNA sequence of their common ancestor, called the *root* of the tree. That sequence is the one the opposing team generated when following Step 2 of Sect. 5.2.

Consider your guess for the first nucleotide of the ancestral DNA sequence. If the first nucleotides of the three sequences of the alignment are A, T, and A, then A would seem like a good guess for their ancestor. You would be more sure of that guess if the first nucleotides were A, A, and A. In other words, your probability that A is the first nucleotide of the ancestral DNA sequence would be greater if the first nucleotides of the *tips* (the sequences of the alignment) were A, A, and A than if they were A, T, and A. That sentence may be more clear when abbreviated like this:

$$\Pr\left(\text{ancestor} = A | \text{tip}_1 = A, \text{tip}_2 = A, \text{tip}_3 = A\right) > \Pr\left(\text{ancestor} = A | \text{tip}_1 = A, \text{tip}_2 = T, \text{tip}_3 = A\right),$$

which is short for "Your probability that the first ancestral nucleotide is A, given that the first nucleotides of the tips are A, A, and A, is greater than your probability that the first ancestral nucleotide is A, given that the first nucleotides of the tips are A, T, and A." (With some practice, the abbreviated forms become much easier to use than the corresponding English sentences.) On the other hand, if the first nucleotides are instead A, T, and G, then you would be even less sure that the ancestral nucleotide at that site (position) is A. In other words,

$$\Pr\left(\text{ancestor} = A | \text{tip}_1 = A, \text{tip}_2 = T, \text{tip}_3 = G\right) < \Pr\left(\text{ancestor} = A | \text{tip}_1 = A, \text{tip}_2 = T, \text{tip}_3 = A\right).$$

In fact, $\Pr\left(\text{ancestor} = A | \text{tip}_1 = A, \text{tip}_2 = T, \text{tip}_3 = G\right)$ might be so small that you would not want to guess at all. In the Substituter game, those probabilities may be calculated by following the steps of Sect. 5.3.

For each alignment of three DNA sequences provided by the opposing team, follow these steps:

(1) At each of the 12 sites of the DNA sequence you are trying to guess:

  (a) Use one of these methods of guessing the ancestral nucleotide at that site:

    (i) Democratic estimation. Like in *Pirates of the Caribbean: The Curse of the Black Pearl* (https://bit.ly/2GteuYn, accessed October 5, 2020), each of the three tip nucleotides at that position votes for itself. Guess the ancestral nucleotide to match the nucleotide with two or three votes from the tips. If there is no such nucleotide (because all three tip nucleotides differ from one another), then do not guess.
    (ii) Bayesian estimation. Follow the steps of Sect. 5.3.
    (iii) Maximum likelihood estimation. Follow the steps of Sect. 5.4.

  (b) Write down one of these symbols:

    (i) To record your guess of the ancestral nucleotide, write it down at that site as "A," "C," "G," or "T."
    (ii) If you do not guess that ancestral nucleotide, then write "?" at that site.

(2) Tell the opposing team your team's estimate of its ancestral DNA sequence. Your team's estimate is the sequence of 12 symbols from the allowed symbols "A," "C," "G," "T," and "?". Your team obtained that sequence by following the above steps.

(3) Ask the opposing team for the actual ancestral DNA sequence.

(4) Compute your team's score by giving your team 1 point for every ancestral nucleotide you guessed correctly and by penalizing your team by 2 points for every ancestral nucleotide you guessed incorrectly. For each "?", you neither gain nor lose points.

5.2.2   Estimating an Ancestral Protein Sequence

This section explains how to apply the rules of Sect. 5.2.1 to estimate an ancestral protein sequence. On the basis of the alignment of the three protein sequences that your opponent generated from the DNA sequences in Step 4 of Sect. 3.1.1.1, estimate their ancestral protein sequence by following democratic estimation in Step 1 for each of the 4 sites of the protein sequence you are trying to guess, except using amino acid symbols instead of nucleotide symbols. That is the sequence of amino acids corresponding to the ancestral DNA sequence according to the standard genetic code. The scoring is the same as described in Step 4 for DNA sequences.

5.3   Probabilities for the First Site of the Ancestral DNA Sequence

How would you compute $\Pr(\text{ancestor} = A | \text{tip}_1 = A, \text{tip}_2 = T, \text{tip}_3 = G)$, a probability mentioned in Sect. 5.2.1? If you knew which expected tree among those in Fig. 3.1 that your opponent used to generate the three tip DNA sequences of the alignment, then you could use *Bayes's theorem*, which says

$$\Pr(\text{ancestor} = A | \text{tip}_1 = A, \text{tip}_2 = T, \text{tip}_3 = G)$$

$$= \frac{\Pr(\text{ancestor} = A) \Pr(\text{tip}_1 = A, \text{tip}_2 = T, \text{tip}_3 = G | \text{ancestor} = A)}{\Pr(\text{tip}_1 = A, \text{tip}_2 = T, \text{tip}_3 = G)}.$$

To calculate that, you would follow these steps to get numbers for the three probabilities on the right-hand side of the formula:

(1) The first probability to determine is $\Pr(\text{ancestor} = A)$. It is called a *prior probability* since you could get it before knowing that $\text{tip}_1 = A, \text{tip}_2 = T, \text{tip}_3 = G$. By contrast, $\Pr(\text{ancestor} = A | \text{tip}_1 = A, \text{tip}_2 = T, \text{tip}_3 = G)$ is called a *posterior probability* since it can only be calculated after observing

that $tip_1 = A$, $tip_2 = T$, $tip_3 = G$. You can determine the numeric value of $\Pr(ancestor = A)$ by reviewing how your opponent generated the ancestor (Step 2 of Sect. 3.1.1.1); see Exercise 2.

(2) The next probability to determine in the formula for Bayes's theorem is $\Pr(tip_1 = A, tip_2 = T, tip_3 = G|ancestor = A)$. It is called the *likelihood* since it is what the probability of observing that $tip_1 = A$, $tip_2 = T$, $tip_3 = G$ would be if A were the ancestral nucleotide. Since the dice rolls in Sect. 3.1.1.1 do not influence each other, we have

$$\Pr(tip_1 = A, tip_2 = T, tip_3 = G|ancestor = A) =$$

$$\Pr(tip_1 = A|ancestor = A) \times \Pr(tip_2 = T|ancestor = A) \times \Pr(tip_3 = G|ancestor = A).$$

Here are the sub-steps for calculating each of the probabilities in the right-hand side of that formula:

(a) Let us focus on getting a number for the $\Pr(tip_1 = A|ancestor = A)$. Assuming for simplicity that there are not multiple substitutions at the first site, $\Pr(tip_1 = A|ancestor = A)$ is equal to the probability that there *is not* a substitution at the first site given that the ancestral nucleotide is A:

$$\Pr(tip_1 = A|ancestor = A) = \Pr(0 \text{ substitutions at first site}|ancestor = A)$$

$$= 1 - \Pr(1 \text{ substitution at first site}|ancestor = A).$$

In that formula, $\Pr(1 \text{ substitution at first site}|ancestor = A)$ is the probability that there is a substitution at the first site. That is approximately equal to the total number of substitutions in the branch from the ancestral sequence to the first tip of the alignment's expected tree of Fig. 3.1 times the probability that the first site is the one experiencing a substitution:

$$\Pr(1 \text{ substitution at first site}|ancestor = A)$$

$$= (\text{total expected branch length from ancestor to } tip_1)$$

$$\times \Pr(1 \text{ substitution at first site}|1 \text{ substitution at a site}).$$

For example, if the expected tree is Tree B in Fig. 3.1, then

$$(\text{total expected branch length from ancestor to } tip_1) = 4$$

since that is 2 from the grandmother to the mother plus 2 from the mother to the first sister. For $\Pr(1 \text{ substitution at first site}|1 \text{ substitution at a site})$, see Exercise 3a. With the numeric values of those probabilities in hand, we can now calculate the numeric value of $\Pr\left(\text{tip}_1 = A|\text{ancestor} = A\right)$:

$$\Pr\left(\text{tip}_1 = A|\text{ancestor} = A\right)$$
$$= 1 - \left(\text{total expected branch length from ancestor to tip}_1\right)$$
$$\times \Pr\left(1 \text{ substitution at first site}|1 \text{ substitution at a site}\right).$$

That is the first of the three probabilities needed in the formula for $\Pr\left(\text{tip}_1 = A, \text{tip}_2 = T, \text{tip}_3 = G|\text{ancestor} = A\right)$.

(b) Next, we need a number for the $\Pr\left(\text{tip}_2 = T|\text{ancestor} = A\right)$ in the formula for Pr. Again assuming that there are not multiple substitutions at the first site, $\Pr\left(\text{tip}_2 = T|\text{ancestor} = A\right)$ is equal to the probability that there *is* a substitution at the first site given that the ancestral nucleotide is A *and* that the substitution is T as opposed to another nucleotide:

$$\Pr\left(\text{tip}_2 = T|\text{ancestor} = A\right) = \Pr\left(1 \text{ substitution at first site}|\text{ancestor} = A\right)$$
$$\times \Pr\left(\text{tip}_2 = T|1 \text{ substitution at first site}\right).$$

Using what we got from sub-step 2a except with tip_2 in place of tip_1, we then have

$$\Pr\left(\text{tip}_2 = T|\text{ancestor} = A\right) = \left(\text{total expected branch length from ancestor to tip}_2\right)$$
$$\times \Pr\left(1 \text{ substitution at first site}|1 \text{ substitution at a site}\right)$$
$$\times \Pr\left(\text{tip}_2 = T|1 \text{ substitution at first site}\right).$$

The total expected branch length from ancestor to tip_2 is all the branch lengths from the grandmother to the second sequence of the alignment added up, using the branch lengths from an expected tree in Fig. 3.1. For $\Pr\left(1 \text{ substitution at first site}|1 \text{ substitution at a site}\right)$, see your answer to Exercise 3a. For $\Pr\left(\text{tip}_2 = T|1 \text{ substitution at first site}\right)$, move on to Exercise 3b.

(c) The calculations of $\Pr\left(\text{tip}_3 = G|\text{ancestor} = A\right)$ are saved for Exercise 3c. That is the last probability needed for calculating $\Pr(\text{tip}_1 = A, \text{tip}_2 = T, \text{tip}_3 = G|\text{ancestor} = A)$.

(3) The last probability to determine in the formula for Bayes's theorem is the one in its denominator: $\Pr\left(\text{tip}_1 = A, \text{tip}_2 = T, \text{tip}_3 = G\right)$. Its value is given by

$$\Pr\left(\text{tip}_1 = A, \text{tip}_2 = T, \text{tip}_3 = G\right) = \Pr(\text{ancestor} = A)\Pr\left(\text{tip}_1 = A, \text{tip}_2 = T, \text{tip}_3 = G|\text{ancestor} = A\right)$$
$$+ \Pr(\text{ancestor} = C)\Pr\left(\text{tip}_1 = A, \text{tip}_2 = T, \text{tip}_3 = G|\text{ancestor} = C\right)$$
$$+ \Pr(\text{ancestor} = G)\Pr\left(\text{tip}_1 = A, \text{tip}_2 = T, \text{tip}_3 = G|\text{ancestor} = G\right)$$
$$+ \Pr(\text{ancestor} = T)\Pr\left(\text{tip}_1 = A, \text{tip}_2 = T, \text{tip}_3 = G|\text{ancestor} = T\right).$$

The good news is you already have the numeric values of $\Pr(\text{ancestor} = A)$ from Step 1 and $\Pr(\text{tip}_1 = A, \text{tip}_2 = T, \text{tip}_3 = G | \text{ancestor} = A)$ from Step 2. The bad news is that to get the other three terms in the right-hand side of the formula for $\Pr(\text{tip}_1 = A, \text{tip}_2 = T, \text{tip}_3 = G)$, you need to follow *both* of those steps for *each* of these:

(a) Ancestor $= C$ instead of ancestor $= A$
(b) Ancestor $= G$ instead of ancestor $= A$
(c) Ancestor $= T$ instead of ancestor $= A$

Plug the numbers you obtained from those steps into the right-hand side of the formula for Bayes's theorem to calculate the numeric value of $\Pr(\text{ancestor} = A | \text{tip}_1 = A, \text{tip}_2 = T, \text{tip}_3 = G)$. That is the posterior probability that the first site in the ancestral sequence is A. Then follow the analogous steps to calculate the numeric value of $\Pr(\text{ancestor} = C | \text{tip}_1 = A, \text{tip}_2 = T, \text{tip}_3 = G)$, $\Pr(\text{ancestor} = G | \text{tip}_1 = A, \text{tip}_2 = T, \text{tip}_3 = G)$, and $\Pr(\text{ancestor} = T | \text{tip}_1 = A, \text{tip}_2 = T, \text{tip}_3 = G)$. To check your work, make sure that all of those posterior probabilities add up to 100%:

$$\Pr(\text{ancestor} = A | \text{tip}_1 = A, \text{tip}_2 = T, \text{tip}_3 = G) + \Pr(\text{ancestor} = C | \text{tip}_1 = A, \text{tip}_2 = T, \text{tip}_3 = G) +$$

$$\Pr(\text{ancestor} = G | \text{tip}_1 = A, \text{tip}_2 = T, \text{tip}_3 = G) + \Pr(\text{ancestor} = T | \text{tip}_1 = A, \text{tip}_2 = T, \text{tip}_3 = G) = 1.$$

If you had to guess the first nucleotide of the ancestral DNA sequence, the best guess would be the one with the highest posterior probability. But you do not need to guess since "?" is an option in the game rules of Sect. 5.2.1. The best move for the scoring system of Step 4 of Sect. 5.2.1 is to guess the nucleotide of the highest posterior probability if it is over 2/3. If it is less than 2/3, then write "?" instead of guessing. If you play Substituter enough times, that strategy would in theory guarantee the highest possible total score for your team.

The estimation method described in this section is *Bayesian* in the sense that it uses Bayes's theorem (Sect. 1.2.3). In scientific practice, Bayesian methods of ancestral tree estimation account for some of the uncertainty about the expected tree and use substitution models that are more realistic than the simulation rules of Sect. 3.1.1.1.

That is all there is to it for estimating the first nucleotide of the ancestral DNA sequence. Now just 11 more estimated nucleotides to go! The next section shows how to take a shortcut if your team does not insist on the theoretically optimal strategy.

5.4   Maximum Likelihood for the First Site of the Ancestral DNA Sequence

Bayesian methods of estimation are criticized not only for being too complicated but also for relying on guessed values of prior probabilities as if they were known to be the true values. A popular alternative method of guessing is called *maximum likelihood estimation*. Some *empirical Bayes methods* combine the two approaches by estimating prior probabilities by maximum likelihood and then plugging them into Bayes's theorem [e.g., 12, chapter 8], whereas other empirical Bayes methods account for the uncertainty in that estimation (Sect. 7.2).

Again assume the first nucleotides of the three sequences of the alignment are A, T, and G. These steps describe how to use maximum likelihood to estimate the first nucleotide of the ancestral DNA sequence that was used by the opposing team to generate those tip sequences:

(1) Compute the four likelihood values:

   (a) Calculate $\Pr\left(\text{tip}_1 = A, \text{tip}_2 = T, \text{tip}_3 = G \mid \text{ancestor} = A\right)$ as described in Step 2 of Sect. 5.3.
   (b) Calculate $\Pr\left(\text{tip}_1 = A, \text{tip}_2 = T, \text{tip}_3 = G \mid \text{ancestor} = C\right)$ the same way except with ancestor $= C$ instead of ancestor $= A$.
   (c) Calculate $\Pr\left(\text{tip}_1 = A, \text{tip}_2 = T, \text{tip}_3 = G \mid \text{ancestor} = G\right)$ the same way except with ancestor $= G$ instead of ancestor $= A$.
   (d) Calculate $\Pr\left(\text{tip}_1 = A, \text{tip}_2 = T, \text{tip}_3 = G \mid \text{ancestor} = T\right)$ the same way except with ancestor $= T$ instead of ancestor $= A$.

(2) Use these rules to determine the symbol to write when performing Step 1 of Sect. 5.2.1:

   (a) If one of the four likelihood values from Step 1 is higher than the other three, then the ancestor corresponding to the highest likelihood is the *maximum likelihood estimate* and is written as your team's guess for the first nucleotide of the ancestral sequence.
   (b) If none of those four likelihood values is higher than the other three, then write "?" rather than guessing the first nucleotide of the ancestral sequence.

Analogous steps would generate the maximum likelihood estimates of the other 11 nucleotides of the ancestral DNA sequence used by the opponent.

Hall [53, chapter 16] explains how to use the MEGA 7 software to do that with real sequences and assuming more realistic models of substitution than the model described in Sect. 3.1.1.1. That explanation also applies to the next version (MEGA X). The main exception is its subheading "NUMBER THE INTERNAL NODES" (pp. 251-253), which may be replaced with the section of Hall [54] entitled, "A much easier way to label tree nodes." For step-by-step instructions, see Exercise 10c of Sect. 5.6.

In spite of its appeal, maximum likelihood estimation tends to be inaccurate when the number of sequences in the alignment is small.

5.5 Uncertainty in Ancestral Sequence Estimates

5.5.1 Sources of Uncertainty in Ancestral Sequence Estimates

How reliable are the posterior probabilities given by the method of Sect. 5.3? Their reliability is compromised to some extent by unrealistic approximations and by the simplicity of the mathematical model underlying Substituter. The probabilities of MEGA X and other software applications that use more realistic models provide more reliable probabilities.

However, even the best methods of calculating posterior probabilities tend to overestimate the probabilities of ancestral nucleotides. That is because no mathematical model can account for all of the uncertainty involved in estimating ancestral sequences. Every method relies at least implicitly on a mathematical model, and it is impossible to construct a mathematical model without making simplifying assumptions.

Specifically, we not only have all of the sources of uncertainty about the tree listed in Sect. 3.4 but also additional uncertainty about the ancestors. For even if we knew the topology of the tree, the ancestors would still be unknown, as seen in Sect. 1.1. The following sources of uncertainty play a large role in uncertainty about any ancestors.

5.5.1.1 Uncertainty About the Statistical Method and About Prior Probabilities

While a Bayesian statistical method (Sect. 5.3) would be ideal if the prior probabilities were known, there is controversy about their use to the extent that they are uncertain (Sect. 4.2.1.1). In some cases, the maximum likelihood statistical method (Sect. 5.4) can perform better.

Some of the uncertainty about which statistical method to use can be represented by noting how changing the method impacts the ancestral nucleotides guessed. In general, the extent to which different methods give different results reflects how much uncertainty there is about those results [53, pp. 177–178]. Torres et al. [117] observed more disagreement between maximum likelihood estimation, Bayesian estimation, and maximum parsimony at the phylum and kingdom levels than at lower taxa. That agrees with the intuition that there would be more uncertainty when the estimated time scales are larger, but there can still be considerable uncertainty even for the lowest taxa (Sect. 5.5.1.2).

In addition, when using Bayesian statistics, some of the uncertainty about the values of the prior probabilities can be represented by noting the degree to which changing them affects the posterior probabilities. Ideally, the posterior distributions would be robust in the sense of being insensitive to the choices of prior distributions. Otherwise, the results are uncertain to at least the extent as the prior distributions.

A mathematical way to account for uncertainty about the prior probabilities is described in Sect. A.3 of Appendix A.

5.5.1.2   Uncertainty About the Model Assumptions

Both the maximum likelihood method and the Bayesian method require model assumptions. The uncertainty about those assumptions needs to be propagated in order to accurately quantify the uncertainty of the results (Sect. 3.4.4).

Although that uncertainty is expected to increase with the time scale (Sect. 5.5.1.1), the uncertainty can be substantial even when working on a time scale of months as opposed to millions of years. For example, Pipes et al. [101], obtaining conflicting results on the phylogenetic tree for variants of the SARS-CoV-2 depending on which assumptions were made, concluded that molecular sequences could not be used to make reliable estimates about its origin without other types of data. Difficulties include distant animal variants and the lack of samples close to the root, motivating a novel maximum likelihood method with uncertainty quantified by a new bootstrap approach [76].

5.5.2   Correcting for Unquantified Uncertainty in Ancestral Sequence Estimates

The sources of uncertainty listed in Sects. 3.4 and 5.5.1 are known to the research community. As a result, many new mathematical models are published every year to incorporate more and more uncertainty into the results.

At the same time, to make inferences about ancestral sequences, we cannot wait for the models and software needed to account for all relevant uncertainty. We have to use the best methods currently available, knowing they tend to report probabilities of ancestral nucleotides that are more confident than warranted given the uncertainty that their models do not represent. One solution is to correct those posterior probabilities for such unquantified uncertainty.

To do that, we need to estimate the extent of the uncertainty that is not represented by the model used in a method. That extent is a number between 0% and 100% and is abbreviated by the letter u, called the *proportion of unquantified uncertainty*, first seen in Sect. 4.2.3. The symbol u stands for the percentage of uncertainty that the mathematical model leaves unrepresented. If the model accounted for 100% of the uncertainty, then $u = 0\%$. At the other extreme, if the model accounts for none of the uncertainty, then $u = 100\%$. In most cases, u should be set between those extremes, as in Exercise 8.

Since u is the probability that the results do not apply, $100\% - u$ is the probability that they do apply and for that reason is multiplied by probabilities in the results (Sects. 4.2.3 and 4.3). In the case of estimating assumed ancestral sequences, the probabilities of ancestral sites are corrected for unquantified uncertainty by multiplying them by $100\% - u$.

For example, the corrected version of a posterior probability such as $\Pr(\text{ancestor} = A | \text{tip}_1 = A, \text{tip}_2 = T, \text{tip}_3 = G)$ is

$$\Pr_u(\text{ancestor} = A | \text{tip}_1 = A, \text{tip}_2 = T, \text{tip}_3 = G) = (1 - u) \times \Pr(\text{ancestor} = A | \text{tip}_1 = A, \text{tip}_2 = T, \text{tip}_3 = G)$$

and is called the *corrected posterior probability*. In the case that 50% of the uncertainty is unquantified by that probability, we plug 0.5 into u, obtaining

$$\Pr_{50\%}(\text{ancestor} = A | \text{tip}_1 = A, \text{tip}_2 = T, \text{tip}_3 = G) = (100\% - 50\%) \times \Pr(\text{ancestor} = A | \text{tip}_1 = A, \text{tip}_2 = T, \text{tip}_3 = G)$$
$$= (1 - 0.50) \times \Pr(\text{ancestor} = A | \text{tip}_1 = A, \text{tip}_2 = T, \text{tip}_3 = G)$$
$$= \frac{\Pr(\text{ancestor} = A | \text{tip}_1 = A, \text{tip}_2 = T, \text{tip}_3 = G)}{2}$$

as the corrected posterior probability when $u = 50\%$. The corrected posterior probability (with u in the subscript of $\Pr_u$) can be used in place of the uncorrected posterior probability (without u in the subscript), when deciding on an estimated ancestral nucleotide, as in Sect. 5.3. Any other posterior probability may be corrected by plugging it into the formula in place of $\Pr(\text{ancestor} = A | \text{tip}_1 = A, \text{tip}_2 = T, \text{tip}_3 = G)$.

You may check for human error by making sure that $\Pr_u(\text{ancestor} = A | \text{tip}_1 = A, \text{tip}_2 = T, \text{tip}_3 = G)$ is lower than $\Pr(\text{ancestor} = A | \text{tip}_1 = A, \text{tip}_2 = T, \text{tip}_3 = G)$. Exercise 7 asks why that should be the case. Exercise 8 provides an opportunity to become comfortable with $\Pr_u(\text{ancestor} | \text{tip}_1, \text{tip}_2, \text{tip}_3)$, the generic form of $\Pr_u(\text{ancestor} = A | \text{tip}_1 = A, \text{tip}_2 = T, \text{tip}_3 = G)$, for various values of u. Which value should be used in practice (Exercise 9)?

5.6 Exercises

(1) Estimate an ancestral protein sequence of Substituter, the dice game introduced in Sect. 3.1:

    (a) Ask the opposing team for a protein alignment of three tip sequences from any of the expected trees in Fig. 3.1, except Tree A.

    (b) For *each* of the 4 sites of the ancestral protein sequence coming from the ancestral DNA sequence as per the standard genetic code:

- According to the democratic estimation method of Sect. 5.2.1, with amino acids in place of nucleotides (Sect. 5.2.2), which symbol is generated as the guess of the amino acid at that site of the ancestral protein sequence?

(c) Ask the opposing team for the root sequence generated from the expected tree.
(d) If the root is a sequence of 12 nucleotides, use the standard genetic code to convert it to a sequence of 4 amino acids.
(e) Compute the total scores achieved by democratic estimation by following Step 4 of Sect. 5.2.1, with amino acids in place of nucleotides (Sect. 5.2.2).

(2) What is the numeric value of $\Pr(\text{ancestor} = A)$ in Step 1 of Sect. 5.3? Hint: think about the probability of getting each outcome of the 4-sided die used by the opposing team when it consulted Table 3.1.
(3) These questions are for Step 2 of Sect. 5.3:

(a) For each of the expected trees in Fig. 3.1, what is the numeric value of

$$\Pr(1 \text{ substitution at first site} \mid 1 \text{ substitution at a site})?$$

In other words, if there is a nucleotide substitution among the 12 nucleotide sites of the DNA sequence, what is the probability that it takes place at the first site as opposed to one of the other 11 sites? Hint: review how the DNA sequence evolved according to rolling the 12-sided die in Step 3 of Sect. 3.1.1.1.
(b) What is the numeric value of $\Pr(\text{tip}_2 = T \mid 1 \text{ substitution at first site})$? Hint: review how the DNA sequence evolved according to rolling the 6-sided die in Step 3 of Sect. 3.1.1.1.
(c) For each of the expected trees in Fig. 3.1, what is the numeric value of $\Pr(\text{tip}_3 = G \mid \text{ancestor} = A)$? Hint: would it be calculated the same way as $\Pr(\text{tip}_1 = A \mid \text{ancestor} = A)$ or the same way as $\Pr(\text{tip}_2 = T \mid \text{ancestor} = A)$?

(4) These questions are for the method of maximum likelihood estimation explained in Sect. 5.4:

(a) For each of the expected trees in Fig. 3.1, what is the maximum likelihood estimate of the nucleotide at the first site, assuming you observed A, T, and G as the first nucleotides of the tip sequences?
(b) Does your answer make sense for Tree A, which actually has three separate root sequences rather than a common ancestor? What would happen if you computed the maximum likelihood estimate for sequences that were not homologous?

(5) (a) What is the difference in biological meaning between the likelihood $\Pr(\text{tip}_1 = A, \text{tip}_2 = T, \text{tip}_3 = G \mid \text{ancestor} = A)$ and the posterior probability

$\Pr\left(\text{ancestor} = A | \text{tip}_1 = A, \text{tip}_2 = T, \text{tip}_3 = G\right)$? (b) Which is harder to compute? (c) Which tends to generate more accurate estimates of the ancestral sequence when the prior probabilities (Step 1 of Sect. 5.3) are known?

(6) Estimate an ancestral DNA sequence of Substituter (Sect. 5.2.1):

(a) Ask the opposing team for a DNA alignment of three tip sequences from any of the expected trees in Fig. 3.1, except Tree A.

(b) For the *first* site of the ancestral DNA sequence from which the alignment was generated:

(i) According to the method of Sect. 5.3, which nucleotide has the highest posterior probability to be the nucleotide?

(ii) Which symbol does that method generate as the guess?

(c) For *each* of the 12 sites of the ancestral DNA sequence from which the alignment was generated:

(i) According to the democratic estimation method of Sect. 5.2.1, which symbol is generated as the guess?

(ii) According to the method of Sect. 5.4, which nucleotides have the highest likelihoods? Which symbol does that method generate as the guess?

(d) Compute the total scores achieved by democratic estimation and by the method of Sect. 5.4.

(i) Which method was more accurate?

(ii) Which method would be more accurate with many DNA sequences that share a common ancestor?

(7) Why should a correction for sources of uncertainty about an ancestral sequence that are not quantified in a model (Sects. 3.4 and 5.5.1) always *reduce* the posterior probability, bringing it closer to 0% (Sect. 5.5.2)? In other words, why should not the uncertainty correction ever *increase* the posterior probability, making it closer to 100%?

(8) According to Sect. 5.5.2's correction of the posterior probabilities computed for Exercise 6b:

(a) How would you answer these questions using $u = 50\%$?

(i) What is your value of $\Pr_{50\%}\left(\text{ancestor} | \text{tip}_1, \text{tip}_2, \text{tip}_3\right)$?

(ii) Using that corrected posterior probability, which symbol is generated as the guess?

(iii) Does it differ from your answer to Exercise 6(b)ii?

(b) How would you answer those questions using $u = 25\%$?

(c) How would you answer those questions using $u = 75\%$?

(d) How would you answer those questions using $u = 100\%$? Is your value of $\Pr_{100\%}(\text{ancestor}|\text{tip}_1, \text{tip}_2, \text{tip}_3)$ what you would expect if the model failed to account for *any* of the relevant uncertainty?

(e) How would you answer those questions using $u = 0\%$? Is your value of $\Pr_{0\%}(\text{ancestor}|\text{tip}_1, \text{tip}_2, \text{tip}_3)$ what you would expect if the model successfully accounted for *all* of the relevant uncertainty?

(9) These foundational questions ask you to give more thought to the values of unquantified uncertainty used in Exercise 8 ($u = 0\%, 25\%, 50\%, 75\%, 100\%$):

(a) In your opinion, which of those values of u would be most appropriate for each time scale, starting from months and then increasing to hundreds of MYs? Hint: keep in mind the definition of the proportion of unquantified uncertainty (u) given in Sect. 5.5.2 while carefully considering the sources of uncertainty mentioned in Sects. 3.4 and 5.5.1. How would you defend your answer?

(b) What sources of uncertainty are not mentioned in Sects. 3.4 and 5.5.1? Hint: take a peek at Chap. 6. How much would those additional sources increase the value of u that you estimated in Exercise 9a?

(10) Follow these steps to apply this chapter's unquantified uncertainty correction to real data:

(a) Install MEGA X after downloading it from https://www.megasoftware.net (accessed August 17, 2021).

(b) On your computer, find and double-click the "Drosophila_Adh.meg" alignment file that came with MEGA X.

(c) Save the estimated ancestral sequences this way [cf. 53, 54]:

   (i) Click "PHYLOGENY" and then "Construct/test Maximum Likelihood Tree..."
   (ii) Click "View," "Show/Hide," and "Node IDs."
   (iii) Click "Ancestors" and then "Show all."
   (iv) Click "Ancestors" and then "Detailed text export."
   (v) Rename the file to "Drosophila.txt."

(d) Drag the file you just created ("Drosophila.txt") to Microsoft Excel, and then save it as an Excel file called "Drosophila.xlsx."

(e) Merge that Excel file with "ancestor uncertainty.xlsx" (Springer's extra supplementary material) in this way:

   (i) Copy columns A–E of the Excel file you created ("Drosophila.xlsx").
   (ii) Paste them into "ancestor uncertainty.xlsx."
   (iii) Save the file as "Drosophila ancestor uncertainty.xlsx."

(f) In that Excel file ("Drosophila ancestor uncertainty.xlsx"), change the unquantified uncertainty parameter to other values between 0 and 1. How does the proportion of unquantified uncertainty affect the probability of each nucleotide for the assumed ancestral sequences?

Chapter 6
Testing Hypotheses of Molecular Evolution

In applying mathematics to subjects such as physics or statistics we make tentative assumptions about the real world which we know are false but which we believe may be useful nonetheless.
– George E. P. Box[1]

Is it of the slightest use to reject a hypothesis until we have some idea of what to put in its place? ... There has not been a single date in the history of the law of gravitation when a modern significance test would not have rejected all laws and left us with no law.

– Sir Harold Jeffreys[2]

In addition to the sources of uncertainty explained in Sects. 3.4, 5.5.1, and 4.2.1, there is uncertainty about whether the neutral theory (Sect. 1.2.2) or one of its alternatives is approximately true or at least adequate as a working hypothesis. Chapter 2 chronicles some challenges faced by the neutral theory.

The main working hypothesis of this chapter was proposed as an alternative to the neutral theory and its nearly neutral variant discussed in Sects. 2.1–2.2. That working hypothesis is summarized in Sect. 6.1. Some potential supporting evidence is then explained in Sect. 6.2. The working hypothesis suggests that the method of phylogenetic tree reconstruction that is outlined in Sect. 6.3. Finally, some open questions are raised in Sect. 6.4.

Electronic Supplementary Material The online version contains supplementary material available at (https://doi.org/10.1007/978-3-031-11958-3_6).

---

[1] "Science and Statistics," *Journal of the American Statistical Association* [25], copyright © American Statistical Association, reprinted by permission of Taylor & Francis Ltd, http://www.tandfonline.com on behalf of American Statistical Association.

[2] *Theory of Probability* (Oxford University Press) [63, §7.22]. Reproduced with permission of the Licensor through PLSclear.

© The Author(s), under exclusive license to Springer Nature Switzerland AG 2022
D. R. Bickel, *Phylogenetic Trees and Molecular Evolution*, SpringerBriefs in Systems Biology, https://doi.org/10.1007/978-3-031-11958-3_6

6.1   Maximum Genetic Diversity Hypothesis

The terminology and concepts of this section follow Huang [59]; see Hu et al.
[58] and Huang [60] for reviews and Wang et al. [124] for a recent development.
The *genetic diversity* of a taxon with respect to a protein or DNA sequence is the
percentage of positions in the sequence that differ among members of the taxon. The
upper limit of that percentage is called the *maximum genetic diversity* of the taxon
and is achieved for rapidly changing sequences. The maximum genetic diversity
measures the fraction of positions in the sequence that are free to change without
negatively impacting the fitness of the members of the taxon. Such changes either
improve fitness or are neutral (Shi Huang, personal communication). The positions
not free to change are considered *conserved*.

The *epigenetic complexity* of a taxon is an average number of cell types of its
individual members. Taxa of higher epigenetic complexity are considered more
complex, whereas those of lower epigenetic complexity are considered simpler.

The maximum genetic diversity hypothesis of Huang [59] makes these claims:

(1) Maximum genetic diversity tends to be higher for simpler taxa and lower for
    more complex taxa. The reason is that the physiology of members of more
    complex taxa relies on more sequence positions, which are for that reason
    conserved, leaving fewer positions free to change.
(2) The positions that are conserved in simpler taxa tend to also be conserved in
    more complex taxa. In other words, the positions that are free to change in
    more complex taxa tend to also be free to change in simpler taxa.
(3) The gradual evolution of sequences takes place at the microevolution level but
    cannot be extrapolated to the scale of macroevolution, as Gould [52] had con-
    cluded largely on the basis of the fossil record. Macroevolution instead involves
    increasing epigenetic complexity and decreasing maximum genetic diversity
    rather than substitutions at positions of a protein or DNA sequence.

    • By contrast, most biologists still consider macroevolution to be an extension
      of microevolution [26, pp. 245, 497].
    • For a brief introduction to microevolution and macroevolution, see Lesk [82,
      p. 6].

6.2   Genetic Equidistance Phenomenon

When sequences of two taxa are compared to each other by computing their
distances to the sequence of a third taxon, the third taxon is called the *baseline
group*. If the baseline group is less related to each of the other taxa than they are
to each other, then it is called an *outgroup*. In terms of Fig. 3.4, an outgroup is a
cousin to the other two taxa, which are sisters of each other. Such comparisons to an

outgroup or to another baseline group are used to test the predictions of hypothesis assuming the neutrality of substitutions.

The maximum genetic diversity hypothesis makes the same predictions as the molecular clock hypothesis (discussed in Sect. 1.2.2 and used in Sect. 3.1.2) except for sequences that evolve rapidly enough that non-conserved positions tend to experience substitutions. For such sequences, Claims 1–2 of the maximum genetic diversity hypothesis (Sect. 6.1) have the consequence of *rapid-evolution saturation*: the estimated number of substitutions between two sequences tends to be equal to the number of non-conserved sites in the simpler of the two taxa due to multiple substitutions at the same sites. That consequence leads to these predictions for substitutions (defined in Sec. 1.2.2) in rapidly evolving sequences [59, Fig. 3]:

(1) **Simpler outgroup.** Considering the tip taxa at the bottom of the tree in Fig. 3.4, suppose Cousin is the simplest and that Sister 1 is the most complex. When the outgroup (Cousin) is simpler than the sister taxa (Sisters 1–2), the distances of the sisters to the outgroup are about equal. In the distance notation of Sect. 3.1.2.1,

$$\overline{(\text{Sister1}) (\text{Cousin})} = \overline{(\text{Sister2}) (\text{Cousin})}.$$

That happens because for rapidly evolving sequences, changes in the lineage leading to the outgroup (Cousin) mask any changes in the lineages leading to the other two taxa (Sisters 1–2). That in turn is a result of rapid-evolution saturation.

- That equality of distances, called the *genetic equidistance phenomenon*, has often been observed and is also a prediction of the molecular clock hypothesis (Fig. 2.1). In fact, the molecular clock hypothesis was proposed in order to explain the genetic equidistance phenomenon (Sect. 2.1).
- The genetic equidistance phenomenon was not predicted by the hypothesis that most differences between the sequences resulted from natural selection. The failure of that selection hypothesis led to several decades of debates related to how much of a role selection played in molecular evolution (Chap. 2).

  - The maximum genetic diversity hypothesis, holding that most differences between the sequences did in fact result from natural selection, explains the genetic equidistance phenomenon as a result of rapid-evolution saturation (Shi Huang, personal communication).

(2) **More complex outgroup.** Relabeling the tips of the tree in Fig. 3.4, suppose Cousin is the most complex and that Sister 1 is the simplest. When the outgroup (Cousin) is more complex than the sister taxa, the distance from the more complex sister (Sister 2) to the outgroup (Cousin) is less than the distance from the simpler sister (Sister 1) to the outgroup (Cousin):

$$\overline{(\text{Sister2}) (\text{Cousin})} < \overline{(\text{Sister1}) (\text{Cousin})}. \tag{6.1}$$

That is because there are fewer conserved sites in the lineage leading to the simpler of the two sisters (Sister 1) and because substitutions in the more complex outgroup (Cousin) are hidden by those of each of the two simpler taxa (Sisters 1-2) due to rapid-evolution saturation.

- That *genetic non-equidistance phenomenon* is not a prediction of the molecular clock hypothesis, which instead predicts that the distance to each of the sister taxa to the outgroup (Cousin) is about equal:

$$\overline{(Sister2)\ (Cousin)} = \overline{(Sister1)\ (Cousin)}. \qquad (6.2)$$

(3) **Simpler non-outgroup baseline group.** Leaving the labels of Fig. 3.4 unchanged, again suppose Cousin is the most complex and that Sister 1 is the simplest. When the baseline group is not an outgroup but is the simplest of the three taxa (Sister 1) and is more closely related to the taxon of intermediate complexity (Sister 2) than to the taxon of highest complexity (Cousin), the distance between the simplest two taxa (Sisters 1-2) is approximately equal to the distance between the simplex taxon (Sister 1) and the most complex taxon (Cousin):

$$\overline{(Sister1)\ (Sister2)} = \overline{(Sister1)\ (Cousin)}. \qquad (6.3)$$

This results from rapid-evolution saturation: substitutions in the lineage leading to the simplest taxon (Sister 1) mask those in the lineages leading to the other two taxa (Sister 2 and Cousin).

- That is not a prediction of the molecular clock hypothesis, which instead predicts that the distance from the simplest taxon (Sister 1) to its sister taxon (Sister 2) is smaller than the distance from the simplest taxon (Sister 1) to the most complex taxon (Cousin):

$$\overline{(Sister1)\ (Sister2)} < \overline{(Sister1)\ (Cousin)}. \qquad (6.4)$$

Example 3 Snakes are simpler than birds, and birds are simpler than humans; snakes and birds are more similar to each other than to humans. For rapidly evolving sequences, the maximum genetic diversity hypothesis then makes these predictions:

- **More complex outgroup.** With the human taxon as the outgroup (Cousin), the distance from birds (Sister 2) to humans is less than the distance from snakes (Sister 1) to humans according to formula (6.1). (The "Cousin" and "Sister" labels in parentheses are those of Fig. 3.4.)

  - The first column of Table 6.1 shows a consistent difference between those distances in the direction predicted by the maximum genetic diversity hypothesis.

That observation indicates that the molecular clock hypothesis does not apply since it predicted those distances to be about equal, as seen in formula (6.2).

- **Simpler non-outgroup baseline group.** With the snake taxon (Sister 1) as a non-outgroup used for reference, the distance of birds (Sister 2) to snakes and the distance of humans (Cousin) to snakes are about equal according to formula (6.3).

  - The second column of Table 6.1 shows a relatively small difference between those distances. That small observed difference is not a prediction of the molecular clock hypothesis, which instead predicts that snakes (Sister 1) would be more consistently closer to birds (Sister 2) than to humans (Cousin) according to formula (6.4). ▲

6.3 Slow Clock Method

If Table 6.1 were used to construct a phylogenetic tree with snakes, birds, and humans, it would say that birds and humans diverged from a common ancestor more recently than the lineage of that ancestor diverged from snakes. In terms of Fig. 3.4, birds and humans are sister taxa related to the more distantly related cousin taxon of snakes, contradicting Example 3. According to Huang [59], the implausibility of those evolutionary relationships casts doubt on phylogenetic trees estimated on the basis of rapidly evolving sequences.

To solve that kind of problem, Huang [59] proposed the *slow clock method* of estimating phylogenetic trees from three taxa:

(1) Only include slowly evolving sequences in the alignment.

   - That is recommended because the resulting estimates are not subject to rapid-evolution saturation and because slow evolution tends to be more neutral and consequently in better agreement with the molecular clock hypothesis (Shi Huang, personal communication).

(2) Use the simplest of the three taxa as the outgroup to construct the distance matrix.

(3) Use a distance-based method of estimating phylogenetic trees.

Example 4 To compare two taxa of pongids (gorillas and chimpanzees) to humans, the slow clock method suggests a simpler outgroup such as orangutans. Tables 6.2 and 6.3 summarize the results. ▲

Table 6.1 Comparison of rapidly evolving protein sequences from snakes (simplest), birds (intermediate complexity), and humans (most complex). The last row displays mean pairwise differences. The displayed 95% confidence intervals of the mean difference in identity assume the 23%-identities of each sample are normally distributed, but the actual uncertainty is higher since only 13 of the 23 proteins were randomly selected. The maximum genetic diversity hypothesis is much better supported in this case than the molecular clock hypothesis (Example 3), though both hypotheses would be rejected at the 5% level according to null hypothesis significance testing [18]. The numbers shown were calculated from the data of Huang [59, Table S3]

Human baseline group	Snake baseline group
Birds more like humans: 23/23	Birds more like snakes: 17/23
Snakes more like humans: 0/23	Humans more like snakes: 6/23
Bird-human identity – snake-human identity = 6.0% ± 1.4%	Bird-snake identity – human-snake identity = 2.6% ± 2.3%

Table 6.2 The numbers of sequences indicating which of the taxa compared (gorillas, humans) is more like the outgroup (orangutans). This information is derived from Huang [59, Table 1]

	Slowly evolving sequences	Rapidly evolving sequences
Gorillas more like orangutans	27	14
Humans more like orangutans	7	16

Table 6.3 The numbers of sequences indicating which of the taxa compared (chimpanzees, humans) is more like the outgroup (orangutans). This information is derived from Huang [59, Table 1]

	Slowly evolving sequences	Rapidly evolving sequences
Chimpanzees more like orangutans	17	8
Humans more like orangutans	3	10

6.4 Questions Raised by Distinguishing Macroevolution from Microevolution

In this chapter, we encountered some of the potential molecular evidence, as opposed to the better known fossil evidence (e.g., Sect. 2.4), that macroevolution is distinct from microevolution. If the gradual process of the neo-Darwinian synthesis cannot be extrapolated to macroevolution, then how should observed differences in epigenetic complexity be explained? What would be the mechanisms for evolution according to molecular versions of the punctuated equilibrium hypothesis (Sects. 2.4–2.7)? Can some kind of epigenetic inheritance [62, chapter 4] fill in the gaps?

Some maverick scientists have concluded that the evidence challenges not only the neo-Darwinian synthesis but also the theory of universal common descent [57, 105, 109]. That, however, is the foundational working hypothesis of molecular phylogenetics, at least as applied to higher level taxa. For without homology in the sense of common ancestry, tree estimation reduces to hierarchical cluster analysis (Sect. 1.2.1).

6.5 Exercises

(1) (a) Why, exactly, does claim 2 of the maximum genetic diversity hypothesis (Sect. 6.1) make the predictions listed in Sect. 6.2? Hint: if there are only a few sites in a sequence that are free to change, there will be more chances for multiple substitutions at the sites that are not conserved according to claim 2. (The concept of multiple substitutions at a site was introduced in Sect. 1.1 and also appeared in Sects. 3.1.2.2 and 3.1.2.3.) (b) Why does the molecular clock hypothesis make the predictions specified in Sect. 6.2?

(2) Why does the maximum genetic diversity hypothesis (Sect. 6.1) suggest working with slowly evolving sequences (Sect. 6.3)? Hint: for slowly evolving
sequences, the maximum genetic diversity hypothesis makes the same predictions as the molecular clock. Why is that?

(3) (a) What part of the title of Huang [59] is supported by the numbers of slowly
evolving sequences in Tables 6.2 and 6.3? (b) Would that part of the title be
supported by the counts of rapidly evolving sequences in Tables 6.2 and 6.3?

(4) Why does the slow clock method of Sect. 6.3 require slowly evolving
sequences? Hint: study the remarks in Sect. 6.3 about Table 6.1, and review
your answers to Exercises 2 and 3.

(5) (a) Based on your understanding of Sect. 6.3, how would you change your
answers to Exercise 5b of Sect. 4.4 and to Exercise 9b of Sect. 5.6? Hint: review
the remarks in Sect. 6.3 about Table 6.1. (b) Does your answer depend on
whether the phylogenetic trees are based on rapidly evolving sequences? Hint:
review your answer to Exercise 4.

(6) (a) How would you start to answer the questions raised in Sect. 6.4? Hint: review
the history of ideas sketched in Sect. 2, keeping in mind its epigraph. (b) Outline
a research project that would address one of those questions.

Chapter 7
Recommendations for Further Reading

> *... a lot of people prefer Bayesian support values to other measures of support, such as the bootstrap ... because they have a likeable tendency to give you higher numbers, making you feel happier about your tree.*
>
> – Lindell Bromham[1]

This chapter is a guide for the readers wanting to delve deeper into some of the topics of previous chapters.

7.1 Molecular Phylogenetics Books

Section 7.1.1 describes books requiring no more mathematics preparation than the present book. Section 7.1.2 briefly comments on books for readers with more knowledge of mathematics.

7.1.1 Phylogenetics Books with Less Mathematics

7.1.1.1 Phylogenetic Trees Made Easy: A How-To Manual [53]

Hall [53] shows step by step how to use phylogenetics software, especially MEGA 7. Nearly, all the MEGA 7 instructions apply with little modification to MEGA X,

---

Electronic Supplementary Material The online version contains supplementary material available at (https://doi.org/10.1007/978-3-031-11958-3_7).

---

[1] *An Introduction to Molecular Evolution and Phylogenetics* (Oxford University Press)[26, p. 431]. © Lindell Bromham 2016. Reproduced with permission of the Licensor through PLSclear.

D. R. Bickel, *Phylogenetic Trees and Molecular Evolution*, SpringerBriefs in Systems Biology, https://doi.org/10.1007/978-3-031-11958-3_7

the version described in Kumar et al. [74] and Stecher et al. [108] and mentioned above in Sects. 3.3, 5.4, and 5.6.

In addition to serving as a software tutorial for the topics introduced above in Chaps. 3–5, Hall [53] exposes the readers to specialized topics such as evolutionary networks and the detection of selection pressures. Like the present book (see the Preface), Hall [53] keeps the mathematics simple, including equations when needed for understanding.

7.1.1.2 An Introduction to Molecular Evolution and Phylogenetics [26]

Bromham [26] places molecular phylogenetics in the context of background topics such as mutation, replication, genomics, genetics, and mechanisms of evolution. Bromham [26] makes heavy use of graphical explanations, much as does Chap. 1, above. Due to its warnings about uncertainty in the output of phylogenetics software, Bromham [26] is cited above in Chaps. 3–4 to motivate their corrections of unquantified uncertainty.

To appeal to a wide audience of biologists, Bromham [26] avoids formulas as a matter of principle. For example, Bromham [26, p. 430] translated Bayes's theorem to English sentences much as Laplace had translated probability formulas into French sentences [78]. The principle is not followed slavishly: Bromham [26, p. 418] resorts to an equation in the discussion of evolution rates.

7.1.2 Phylogenetics Books with More Mathematics

Yang [133] gives details of many statistical methods of analyzing sequence data for molecular phylogenetics. The equations and notation are complex enough for describing the methods without recourse to the idealized simplifications seen in Chaps. 3 and 5 above. The author nonetheless made special efforts to make the book accessible to biologists [133, Preface]. Many uncertainties involved in statistical inferences about molecular evolution are thoroughly discussed in Yang [133, chapters 10–11]. The emphasis is on maximum likelihood estimation and Bayesian inference.

Drummond and Bouckaert [42] specifically focus on Bayesian inference about molecular evolution. The authors are the leading developers of BEAST 2 [24], which is currently the most popular software suite dedicated to Bayesian phylogenetics. The work Drummond and Bouckaert [42] is organized into three parts:

(1) The "Theory" part, like much of Yang [133], will appeal to the readers comfortable with calculus, linear algebra, and mathematical notation.
(2) The "Practice" part assists biologists with the use of BEAST 2 without requiring knowledge of the more technical parts.

(3) The "Programming" part, going under the hood of BEAST 2, will interest the readers with coding skills.

An earlier guide to BEAST is chapter 18 in Salemi et al. [104], a book written by the experts in specific areas of molecular phylogenetics, including the preliminary steps of finding and aligning sequences. This chapter describes MrBayes, another software tool for Bayesian inference.

The work Nei and Kumar [92] is written by the creators of the MEGA software mentioned in Sect. 7.1.1.1. Clearly explaining many statistical tests and bootstrap methods, it remains widely cited. Another classic text is Felsenstein [46].

Chapters 13 and 14 of Ewens and Grant [43] describe statistical methods of extracting information on molecular evolution from biological sequence data. Those chapters complement Nei and Kumar [92] in large part by providing a more concise treatment of the topics. Previous chapters of Ewens and Grant [43] give an overview of other statistical methods of analyzing DNA and protein sequences, with an emphasis on the statistics behind BLAST theory. Those methods are used to select and align sequences before the tree estimation methods can be applied. For practical guidance in that use of BLAST, see Hall [53], the book recommended in Sect. 7.1.1.1.

Xia [132] explains much of the mathematics involved in methods of phylogenetic trees reconstructed from sequence data, with an emphasis on distance-based methods and maximum likelihood estimation. The work Xia [132, chapter 2] is cited above in Sect. 3.4.1 on alignment as a source of uncertainty.

7.2 Bioinformatics and Genomics Books

The introductory book by Lesk [82] sets molecular phylogenetics in the context of other methods of computational biology. It is cited in Sect. 3.1.2.1 on distance-based estimation. Lesk [82] displays and discusses the three sequences behind Fig. 3.7.

Abu-Jamous et al. [1] describe many methods of cluster analysis in the context of bioinformatics problems. Distance-based tree estimation, while mathematically a form of hierarchical cluster analysis, has an evolutionary interpretation when homology is accepted as a working hypothesis (Sect. 1.2.1).

Using dice games, Bickel [12] explains statistical methods of analyzing data from genome-wide association studies and from measurements of gene expression and related proteomics and metabolomics data. The empirical Bayes tools (mentioned in Sect. 5.4) primarily apply to simultaneously testing multiple hypotheses. Multiple testing occurs not only in the types of data used in that book but also in an adjusted bootstrap proportion [102] (cf. Sect. 4.1) and also more generally, as seen in Sect. 2.5. Empirical Bayes methods are designed to guard against false positives like those leading to the replication crisis in many fields of science [see 8]. Motivated by that problem, chapter 7 of Bickel [12] explains how such methods scale down to testing a single hypothesis. The book uses a result from the use of confidence theory to propagate the uncertainty in estimating prior distributions [17].

7.3   Imprecise Probability Books

You may recall that Sects. 4.2.3, 4.3, and 5.5.2 explain how to correct a confidence level or posterior probability by multiplying it by the estimated probability that the prior distributions and other model assumptions are adequate. That correction factor, the proportion of uncertainty quantified by the models, is $100\% - u$, where u is the proportion unquantified uncertainty [19].

Strict Bayesians would raise an objection: if all the probabilities are multiplied by $100\% - u$, then the total probability is $100\% - u$ instead of 100%. True, but "it's not a bug, it's a feature" [31], for such estimates honestly reflect the extent of unquantified uncertainty. (A less conservative method [15] is summarized above in Exercise 6b of Sect. 4.4.)

While the corrected probability function cannot be a probability distribution, it qualifies mathematically a *lower probability* function in the theory of imprecise probability. The value of such a function may be interpreted as the sufficiency of the evidence (Appendix A) or as a lower bound on standard probability. Under the latter interpretation, u serves as a "discounting coefficient" in the linear-vacuous [122, §2.9.2] or ε-contamination model of uncertainty described by Augustin et al. [6, §4.7]. Discounting coefficients that do not generate lower probabilities have also been considered [13, 14].

With that in mind, these books on imprecise probability theory are recommended to statisticians and scientists not averse to theorems:

- The work Augustin et al. [6], a unified collection of chapters by various experts, is mentioned out of chronological order since it provides an accessible entry to the topic.
- Walley [122] launched the field, providing mathematical and intuitive arguments for representing uncertainty with imprecise probability.

  - Walley [123] presents later developments by the same author.

- The work Troffaes and de Cooman [118] includes many of the theorems in Walley [122] as well as more recent results.
- 2021 saw the publication of two books with unique perspectives on imprecise probability:

  - Cuzzolin [37] not only presents two decades of research on a geometric approach but also reviews all major flavors of imprecise probability, citing over 2000 works. Discounting is discussed briefly [37, §4.3.6].
  - Weirich [126] puts emphasis on utility functions.

7.4   Power Law Books

Power laws form the core of the big-picture models of self-similar fluctuations seen in Appendices B-C. Such models have been used to describe fluctuating rates of molecular evolution (Sect. 2.8). The recommended starting point is Taleb [111], which offers a lively introduction to power laws.

Lowen and Teich [84] use fractal stochastic processes to define point processes in order to model count data such as the substitutions plotted in Fig. 1.1. A special case of such point processes is the class of intermittent point processes of Appendix B, below. Appendix C provides examples of other cases (Sect. C.2).

West et al. [129] introduce much of the mathematical modeling sketched in Sect. C.1.

Appendix A
Probability Interpretation:
Support by Data and Model Evidence

Some methods of this book bring together concepts normally kept separate within their own statistical theories. As a result, you might be left wondering which interpretation of "probability" is being used. Is it an idealized relative frequency, as in traditional frequentist inference? Is it an idealized level of belief, as in traditional Bayesian inference? Or is it something else entirely? The short answer is that the probability of a hypothesis is the extent to which it is supported by the evidence-considered broadly enough to consider not only the data but also the model specifications as evidence. The long answer is mathematical: this appendix's measure-theoretic framework for interpreting probabilities that may be calculated from standard statistical methods, Bayesian methods, or blended methods.

A.1 An Evidential Generalization of Frequentist and Bayesian Probability

Some methods of this book bring together concepts normally kept separate within their own statistical theories. As a result, you might be left wondering which interpretation of "probability" is being used. Is it an idealized relative frequency, as in traditional frequentist inference (Sect. 1.2.3)? Is it an idealized level of belief, as in traditional Bayesian inference? Or is it something else entirely?

   The short answer is that the probability of a hypothesis is the extent to which it is supported by the evidence—considered broadly enough to consider not only the data but also the model specifications as evidence. The long answer is mathematical: the following measure-theoretic framework for interpreting probabilities that may be calculated from standard statistical methods, Bayesian methods, or blended methods.

D. R. Bickel, *Phylogenetic Trees and Molecular Evolution*, SpringerBriefs in Systems Biology, https://doi.org/10.1007/978-3-031-11958-3

A.2   Evidential Models and Evidential Support Distributions

Let ξ denote an unknown quantity of interest, and let $\varXi$ be the set of possible values of ξ. For example, ξ could be a divergence time or the value of a *prediction*, a measurement to be made in the future. A measurable set of subsets of $\varXi$ is denoted by $\mathfrak{F}$. Let x denote a data set such as a sequence alignment paired with estimates of fossil dates. That x is in a set $\mathscr{X}$ of possible data sets. A function $M\,(\bullet;\,\bullet)$ on the Cartesian product $\mathfrak{F} \times \mathscr{X}$ is called a *general evidential model* if, for every $x \in \mathscr{X}$, $M\,(\bullet;\,x)$ is a probability distribution on the measure space $(\varXi, \mathfrak{F})$. Given a known data set $x \in \mathscr{X}$, $M\,(\bullet;\,x)$ is the *general evidential support distribution* given the pair (x, M), which is called the *general body of evidence*.

The word "general" occurs in the last three terms because they generalize the "evidential model," "evidential support distribution," and "body of evidence" as defined in Bickel [17]. Those special cases may be designated by *special evidential model*, *special evidential support distribution*, and *special body of evidence* in order to distinguish them from the above definitions. The main difference is that a *special* evidential support distribution must be a Bayesian posterior distribution, a frequentist posterior distribution called an "approximate confidence distribution," or a distribution ultimately derived from one or more Bayesian posterior distributions and/or one or more approximate confidence distributions.

By contrast, a *general* evidential support distribution may be a predictive distribution based on a machine learning algorithm that need not satisfy the conditions specified by Bickel [17]. That agrees with the legal precedent of admitting machine learning results as evidence [94, 98] in essentially the same way that prior probabilities are admitted as evidence [70]. The generalization is biologically relevant: deep learning methods, especially those based on deep neural networks, are increasingly used to make estimates or predictions related to molecular evolution [120] and other areas of bioinformatics [61].

A.3   Imprecision Due to Uncertainty

This section extends the framework of Sect. A.2 by using imprecise probability to represent uncertainty about the general evidential models. Detailed background information on imprecise probability theory is given in the books recommended in Sect. 7.3.

A set $\mathscr{M}$ is a *total evidential model* if it is a set of general evidential models on $\mathfrak{F} \times \mathscr{X}$. For molecular phylogenetics, much of the imprecision (inclusion of multiple evidential models) could result from uncertainty about the evolution models (Sect. 5.5.1.1) and prior distributions (Sect. 3.4.4) that underlie the general evidential models. Given a known data set $x \in \mathscr{X}$, the pair $(x, \mathscr{M})$ is called the *total body of evidence*.

Similarly, a set

$$\mathcal{M}(x) = \{M(\bullet; x) : M \in \mathcal{M}\}$$

is the *total evidential support distribution* given $(x, \mathcal{M})$. It follows that $\mathcal{M}(x)$ is a set of general evidential support distributions. Its imprecision comes from the imprecision of $\mathcal{M}$.

Such distributions are most easily understood by considering a concrete hypothesis such as the proposition that the divergence time of two virus strains is between 6 and 12 months ago. For any set $\mathcal{H}$ that is a measurable subset of Ξ, the proposition that $\xi \in \mathcal{H}$ is called a *hypothesis*. The *total evidential support* for the hypothesis that $\xi \in \mathcal{H}$, given $(x, \mathcal{M})$ as the body of evidence, is the set

$$\mathcal{M}(\mathcal{H}; x) = \{M(\mathcal{H}; x) : M \in \mathcal{M}\},$$

which is usually the interval between $\inf \mathcal{M}(\mathcal{H}; x)$ as the lower limit and $\sup \mathcal{M}(\mathcal{H}; x)$ as the upper limit. $\mathcal{M}(\mathcal{H}; x)$ is an example of what is meant by "imprecise probability" (Sect. 7.3).

A.4   Evidential Sufficiency Distributions

Is there enough evidence to conclude that the divergence time of two virus strains is between 6 and 12 months ago? That is the kind of question answered by the tools of this section. The answer depends not only on the evidence but also on the *standard of proof*, which is a probability threshold representing the minimal amount of evidential support that must be accumulated to draw a conclusion. In some legal systems, for example, the standard of proof for a criminal trial is much higher than for a civil trial [106].

Consider s, a number between 50% and 100%, as the standard of proof. The reason s must be at least 50% is that it would not be meaningful to consider a statement proven or to conclude that it is true if it does not achieve at least 50% probability [2].

Given a total body of evidence $(x, \mathcal{M})$, we may conclude that the hypothesis that $\xi \in \mathcal{H}$ is proven if its general evidential support meets or exceeds the standard of proof for every general evidential model in $\mathcal{M}$. Mathematically, the hypothesis that $\xi \in \mathcal{H}$ is called *proven* at standard of proof s if and only if

$$M(\mathcal{H}; x) \geq s$$

for every $M \in \mathcal{M}$.

That can be simplified by combining a mathematical concept with a legal concept. The mathematical concept is what is called a "lower envelope" in imprecise probability theory (Sect. 7.3). The legal concept is the "sufficiency of the evidence," the extent to which there is enough evidence to draw a conclusion [16, 17, 65]. For any $\mathcal{H} \in \mathfrak{F}$, the *evidential sufficiency* of the hypothesis that $\xi \in \mathcal{H}$ is its minimal evidential support in the sense that

$$S(\mathcal{H}; x) = \inf \mathcal{M}(\mathcal{H}; x)$$

given a total body of evidence $(x, \mathcal{M})$. The *evidential sufficiency distribution* given a total body of evidence $(x, \mathcal{M})$ is the function $S(\bullet; x)$ such that $S(\mathcal{H}; x)$ is the evidential sufficiency of any $\mathcal{H} \in \mathfrak{F}$. As a result, the hypothesis that $\xi \in \mathcal{H}$ is proven at standard of proof s if and only if $S(\mathcal{H}; x) \geq s$.

Since the sum of the evidential sufficiency of the hypothesis that $\xi \in \mathcal{H}$ and the evidential sufficiency of the hypothesis that $\xi \notin \mathcal{H}$ can be less than 100%, the evidential sufficiency distribution is not a probability distribution. It is instead what is called a "lower probability function" in Sect. 7.3.

A.5   Special Case: Unquantified Uncertainty Corrected

As noted in Sect. 7.3, the framework of this appendix enables interpreting the uncertainty-corrected confidence levels and probabilities of Sects. 4.2.3, 4.3, and 5.5.2 as levels of evidential sufficiency.

To see that, suppose only a single evidential model M is specified in enough detail to get output from software but that the probability that it fails to adequately quantify the uncertainty is u, the proportion unquantified uncertainty [19]. In this case, the total evidential support for the hypothesis that $\xi \in \mathcal{H}$ is

$$\mathcal{M}(\mathcal{H}; x) = \{(1 - u) M(\mathcal{H}; x) + u P : 0 \leq P \leq 1\}$$
$$= [(1 - u) M(\mathcal{H}; x), (1 - u) M(\mathcal{H}; x) + u]$$
$$= (1 - u) M(\mathcal{H}; x) + \frac{u}{2} \pm \frac{u}{2}.$$

The lowest possible value of that is the evidential sufficiency of the hypothesis that $\xi \in \mathcal{H}$:

$$S(\mathcal{H}; x) = \inf \mathcal{M}(\mathcal{H}; x) = (1 - u) M(\mathcal{H}; x) + \frac{u}{2} - \frac{u}{2} = (1 - u) M(\mathcal{H}; x).$$

In short,

$$S(\mathcal{H}; x) = (1 - u) M(\mathcal{H}; x).$$

Whereas $M(\mathcal{H}; x)$ is the probability that fails to account for u of the uncertainty, $S(\mathcal{H}; x)$ is the uncertainty-corrected probability in terms of Sects. 4.2.3, 4.3, and 5.5.2.

Example 5 Suppose that the 95% confidence interval of the divergence time is between 6 and 12 months ago according to a special evidential model M that leaves 25% of the uncertainty unquantified. Then we plug M (9 ± 3 months; x) = 95% and $u = 25\%$ into the $S(\mathcal{H}; x)$ formula, yielding

$$S(9 \pm 3 \text{ months}; x) = (1 - 0.25) \times 0.95 = 71.25\%.$$

That level of evidential sufficiency is called the "corrected confidence level" in Sect. 4.2.3. ▲

Appendix B
Intermittent Point Processes

Some mathematical details behind the measure of intermittency used in the book are given in this appendix. Its models are called "fractal point processes."

Some mathematical details behind the measure of intermittency used in Fig. 1.1 and explained in Sect. 2.8 are given here. More details can be found in the books mentioned in Sect. 7.4. The models in this appendix are called "fractal point processes" in Sect. 2.8.2.

B.1 Basics of Intermittency

Roughly speaking, the intermittency parameter measures the propensity of the counting process to suddenly jump away from typical values.

Let $N(T)$ be the *counting process* associated with a stationary point process; a realization of $N(T)$ is the number of events that occur within a time period T. It has a distribution that depends on the underlying point process and on the length of the counting time. In particular, we can examine how the second moment of the distribution varies with T. If the underlying point process is scale-invariant, then

$$\left\langle N^2(T) \right\rangle \sim T^{2-C_2}, \tag{B.1}$$

where C_2 is the *correlation codimension*, a measure of the intermittency of the point process on a scale from 0, no intermittency, to 1, completely intermittent [10]. The angular brackets $\langle \bullet \rangle$ denote expectation values known as *ensemble averages* in the physics literature.

The second moment of the counting process can be estimated from the series of the observed counts as follows. Let Z_i denote the number of events in the ith of n consecutive time windows, each of duration T_0, called the *counting time*. Then

© The Author(s), under exclusive license to Springer Nature Switzerland AG 2022
D. R. Bickel, *Phylogenetic Trees and Molecular Evolution*, SpringerBriefs in
Systems Biology, https://doi.org/10.1007/978-3-031-11958-3

the sequence $Z_1, Z_2, \ldots, Z_n$ is called the *count time series*. Construct the *counting process* as the cumulative sum of the count time series:

$$N_k = \sum_{i=1}^{k} Z_i, \; k = 1, 2, \ldots, n. \tag{B.2}$$

Then, for $k = 1, 2, 4, 8, \ldots, k_{max} \leq n$, set

$$S_k = \frac{1}{n - k + 1} \sum_{j=1}^{n-k+1} \left(N_{j+k} - N_j \right)^2. \tag{B.3}$$

S_k is an estimate of $\langle N^2 (kT_0) \rangle$; for point processes with scale-invariant behavior, a log–log plot of kT_0 versus S_k will have a linear region with slope $2 - C_2$.

B.2   Hurst Exponent

Many statistics of fractal and fractal-rate point processes, such as the count autocorrelation function, the interval autocorrelation function, and the interval spectrum show power law scaling; the scaling exponents of these statistics are all related by simple geometric equations. As such, they all equally well characterize the fractal nature of a given process (although practically speaking, some may be easier to estimate than others). Therefore we choose one, the Hurst exponent, as the basis for further discussion.

The *Fano factor*, also called the "index of dispersion," is the ratio of a random variable's variance to its mean. For fractal and fractal-rate processes, the Fano factor of the counting process $N(T)$ scales as

$$F(T) = \frac{\langle N^2 (T) \rangle - \langle N (T) \rangle^2}{\langle N (T) \rangle} \sim T^{(2H-1)}, \tag{B.4}$$

where H, a number between 0 and 1, is called the *Hurst exponent* if the rate process is fractional Gaussian noise Lowen and Teich [84]. The equation for the Fano factor involves the term $\langle N^2 (T) \rangle$, so there is a relationship between H and C_2, but unlike with other scaling exponents, the relationship is not simple, as we demonstrate below. Following Bickel [10], define ρ to be the mean rate of the process, so that $\langle N (T) \rangle = \rho T$. Let T_{max} represent the largest time for which the scaling relationship of Eq. (B.4) holds. Then,

$$F(T) = \frac{\langle N^2(T_{\max}) \rangle \left(\frac{T}{T_{\max}} \right)^{2-C_2} - (\rho T)^2}{\rho T} \times \frac{\left(\frac{T_{\max}}{T} \right)}{\left(\frac{T_{\max}}{T} \right)} \tag{B.5}$$

$$F(T) = \frac{\langle N^2(T_{\max}) \rangle \left(\frac{T}{T_{\max}} \right)^{1-C_2} - \langle N(T_{\max}) \rangle^2 \left(\frac{T}{T_{\max}} \right)}{\langle N(T_{\max}) \rangle}. \tag{B.6}$$

If the process is intermittent (implying $\langle N^2(T_{\max}) \rangle \geq \langle N(T_{\max}) \rangle^2$) and if $T \ll T_{\max}$, then the intermittency has a simple relationship with the Hurst exponent. In that case, the first term in the numerator of Eq. (B.6) dominates the second, and the relation $H = (2 - C_2)/2$ holds. The relation does not hold if T is on the order of $T_{\max}$, nor does it hold for non-intermittent processes, which can have Hurst exponents less than 1 even if $C_2 \approx 0$.

This algorithm is a simple way to estimate the H from estimates of the Fano factor as a function of the counting time:

1. Aggregate the count time series by adding up the original daily count time series within bins of width T.
2. Estimate each $F(T)$ by $\widehat{F}(T)$, the sample variance divided by the sample mean of the aggregated time series.
3. On a log–log plot of the estimated Fano factor scaling exponent as a function T, the linear region will have a slope of $2\widehat{H} - 1$. Solving that for $\widehat{H}$ gives an estimate of H.

B.3   Fractal Renewal Processes

The *fractal renewal processes* [21, 84] are intermittent point processes that are useful for initial modeling and estimation since they have only three parameters. Each fractal renewal process has a probability density of times τ between events proportional to $\tau^{-(1+\beta)}$ in the tails, leading to an intermittency of $C_2 = 1 - \beta$ [10]. Although intermittent, the fractal renewal processes do not have power law distributions of counts [84]. Events from two fractal renewal processes are plotted in Fig. 1.1.

Appendix C
Poisson Processes with Lévy-Stable Substitution Rates

Some technicalities missing from earlier parts of the book are filled in by this appendix. It gives an overview of Lévy-stable processes for substitution rates and point process models for molecular evolution.

Some technicalities missing from Sect. 2.8.3.3 are given here and in the books of Sect. 7.4.

C.1 Lévy-Stable Processes for Substitution Rates

Each *Lévy-stable distribution* or *continuous-stable distribution* is defined by its characteristic function as parameterized by its scale $a > 0$, asymmetry $\beta \in [-1, 1]$, and characteristic exponent $\nu \in (0, 2]$. A *continuous-stable process* is a stochastic process with increments following a continuous-stable distribution.

Lee et al. [81] defined the *discrete-stable process* as a doubly stochastic Poisson process with the rate given by the increments of a continuous-stable process. The Lévy-stable parameter values satisfy $\beta = 1$ and $0 < \nu < 1$ to ensure nonnegativity of the rate, and the probability mass function of the number x of counts decays approximately as the power law $x^{-(1+\nu)}$ in the large-x tail [81]. Thus, the stability of the continuous rate process ensures that, for sufficiently large x, the distribution of counts is approximately self-similar in the sense that its shape is approximately invariant to the counting time.

Mantegna and Stanley [87] demonstrated that a symmetric, sharply truncated continuous-stable process exhibits fractal scaling for a wide range of time scales but, due to its finite variance, eventually converges to a normal distribution. Nakao [91] and Terdik et al. [116] derived analytic expressions for moments of the more tractable smoothly truncated (*ST*) continuous-stable processes [71].

West et al. [129, p. 206] described autocorrelations without a characteristic time scale in increments of a stochastic process $X(t)$ via the $(1/2 + H)$th fractional

© The Author(s), under exclusive license to Springer Nature Switzerland AG 2022 95
D. R. Bickel, *Phylogenetic Trees and Molecular Evolution*, SpringerBriefs in Systems Biology, https://doi.org/10.1007/978-3-031-11958-3

derivative $D_t^{1/2+H} X(t) = \xi(t)$, where $\xi(t)$ is a delta-correlated noise process and $0 < H < 1$, with the degenerate case $H = 1/2$ recovering the independence of increments [86]. The fractional-motion process $X(t)$ is called an *infinite-range (IR) continuous process* since the integral of its autocorrelation function diverges.

C.2   Point Process Models for Molecular Evolution

C.2.1   Poisson-Stable Processes

This section extends to the concept of the discrete-stable processes that model molecular evolution as intermittent. The *Poisson-stable process* generalizes the discrete-stable process of Sect. C.1 in the sense that the former encompasses two-sided continuous-stable processes as well as one-sided continuous-stable processes, as specified below.

The continuous process $X(t)$ underlying that Poisson process may be simultaneously an ST continuous-stable process and an IR continuous process defined in Sect. C.1, as follows. The fractional derivative operator $D_t^{1/2+H}$ ensures that $X(t)$ has long-time memory [129, p. 206], whereas $L(t)$, now the noise process of a one-sided (nondecreasing) ST continuous-stable process, entails that the probability mass function of each increment $X(t'') - X(t')$ decays approximately as the power law $\left(X(t'') - X(t')\right)^{-(1+\nu)}$ in the *tail*, the region of sufficiently high $t'' - t' > 0$, until crossing over toward the normal distribution due to the truncation [71, 77, 116]. The increments of that $X(t)$, the *STIR continuous-stable process*, are the mean numbers of counts of the discrete-stable process. In addition, a strongly nonstationary event time series could be modeled using the increments of a two-sided continuous-stable process $X(t)$ as stationary increments of the nonstationary continuous mean, still called a STIR continuous-stable process. Thus, the probability that x events occur after any time t' and before any greater time t'' is

$$P_T(x) = \frac{1}{x!} \int_0^\infty \mu^x e^{-\mu} p_T(\mu)\, d\mu, \qquad (C.1)$$

where $T = t'' - t'$. In the one-sided case, p_T is the probability density function of $X(T) - X(0)$, which due to stationary is distributed as the continuous increment $X(t'') - X(t')$, whereas in the two-sided case, p_T is the probability density function of $X(T)$, constrained to be nonnegative by the methods of Lowen and Teich [84]. Any stochastic process thereby defined by Eq. (C.1) with $X(t)$ as a STIR continuous-stable process is called a *STIR Poisson-stable process*.

The special subclasses of interest are that of the *ST Poisson-stable process* and that of the *IR Poisson-stable process* defined through Eq. (C.1) with the ST or the IR continuous-stable process (Sect. C.1), respectively, describing the stochastic mean of the Poisson process. While the full generality of the STIR Poisson-stable process may sometimes need to be relaxed when applied to relatively short time series to avoid overfitting the parameters, the full model has advantages for longer time series. The potential advantage of explicitly incorporating smooth truncation into the model (the "ST" in "STIR") is that in real-world systems as opposed to purely mathematical constructs, there are always upper and lower limits to statistical scaling [135].

C.2.2   Poisson-Fractal Processes

Poisson-fractal processes are defined via Eq. (C.1) with means given by *continuous-fractal processes*, understood here as processes with long-range (power law) dependence but without approximating a non-Gaussian continuous-stable process. The class of Poisson-fractal processes includes the *fractal-Gaussian-process-driven Poisson processes* (FGPDPPs) of Lowen and Teich [84], which are characterized by a Hurst-like scaling exponent H that describes the long-range correlations of the underlying fractional Gaussian noise or fractional Brownian motion that takes the place of a STIR continuous-stable process in Eq. (C.1). As the limits of other important fractal point processes, FGPDPPs have enjoyed success in a wide variety of applications [84].

References

1. Abu-Jamous, B., R. Fa, and A.K. Nandi. 2015. *Integrative Cluster Analysis in Bioinformatics.* West Sussex: John Wiley & Sons.
2. Achinstein, P. 2001. *The Book of Evidence.* Oxford: Oxford University Press.
3. Alberdi, A., and M. Gilbert. 2019. A guide to the application of hill numbers to DNA-based diversity analyses. *Molecular Ecology Resources* 19: 804–817.
4. Álvarez-Carretero, S., A. Goswami, Z. Yang, and M. Dos Reis. 2019. Bayesian estimation of species divergence times using correlated quantitative characters. *Systematic Biology* 68: 967–986.
5. Aris-Brosou, S., and Z. Yang. 2003. Bayesian models of episodic evolution support a late Precambrian explosive diversification of the Metazoa. *Molecular Biology and Evolution* 20: 1947–1954.
6. Augustin, T., F. Coolen, G. de Cooman, and M. Troffaes, eds. 2014. *Introduction to Imprecise Probabilities.* Wiley Series in Probability and Statistics. Hoboken: Wiley.
7. Battistuzzi, F.U., Q. Tao, L. Jones, K. Tamura, and S. Kumar. 2018. RelTime Relaxes the Strict Molecular Clock throughout the Phylogeny. *Genome Biology and Evolution* 10: 1631–1636.
8. Bausell, R.B. 2021. *The Problem with Science: The Reproducibility Crisis and What to Do About It.* Oxford: Oxford University Press.
9. Bayer, O. 2012. *A Contemporary in Dissent: Johann Georg Hamann as Radical Enlightener.* Grand Rapids: Wm. B. Eerdmans Publishing.
10. Bickel, D.R. 1999. Estimating the intermittency of point processes with applications to human activity and viral DNA. *Physica A: Statistical Mechanics and Its Applications* 265: 634–648.
11. Bickel, D.R. 2000. Implications of fluctuations in substitution rates: Impact on the uncertainty of branch lengths and on relative-rate tests. *Journal of Molecular Evolution* 50: 381–390.
12. Bickel, D.R. 2019. *Genomics Data Analysis: False Discovery Rates and Empirical Bayes Methods.* New York: Chapman and Hall/CRC. https://davidbickel.com/genomics/.
13. Bickel, D.R. 2020. Departing from Bayesian inference toward minimaxity to the extent that the posterior distribution is unreliable. *Statistics & Probability Letters* 164: 108802.
14. Bickel, D.R. 2021a. Moderating probability distributions for unrepresented uncertainty: Application to sentiment analysis via deep learning. *Communications in Statistics - Theory and Methods.* https://doi.org/10.1080/03610926.2020.1863988.
15. Bickel, D.R. 2021b. Propagating uncertainty about molecular evolution models and prior distributions to phylogenetic trees. Working paper. https://doi.org/10.5281/zenodo.5810696.
16. Bickel, D.R. 2021c. The sufficiency of the evidence, the relevancy of the evidence, and quantifying both with a single number. *Statistical Methods & Applications* 30: 1157–1174.

17. Bickel, D.R. 2022a. Confidence distributions and empirical Bayes posterior distributions unified as distributions of evidential support. *Communications in Statistics - Theory and Methods* 51: 3142–3163.

18. Bickel, D.R. 2022b. Fisher's disjunction as the principle vindicating p-values, confidence intervals, and their generalizations: A frequentist semantics for possibility theory. Working paper. https://doi.org/10.5281/zenodo.6590672.

19. Bickel, D.R. 2022c. Propagating clade and model uncertainty to confidence intervals of divergence times and branch lengths. *Molecular Phylogenetics and Evolution* 167: 107357.

20. Bickel, D.R., and B.J. West. 1998a. Molecular evolution modeled as a fractal Poisson process in agreement with mammalian sequence comparisons. *Molecular Biology and Evolution* 15: 967–977.

21. Bickel, D.R., and B.J. West. 1998b. Molecular evolution modeled as a fractal renewal point process in agreement with the dispersion of substitutions in mammalian genes. *Journal of Molecular Evolution* 47: 551–556.

22. Bickel, D.R., and B.J. West. 1998c. Multiplicative and fractal process in DNA evolution. *Fractals* 6: 211–217.

23. Bleidorn, C. 2017. *Phylogenomics: An Introduction*. New York: Springer International Publishing.

24. Bouckaert, R., J. Heled, D. Kühnert, T. Vaughan, C.H. Wu, D. Xie, M.A. Suchard, A. Rambaut, and A.J. Drummond. 2014. BEAST 2: A software platform for Bayesian evolutionary analysis. *PLoS Computational Biology* 10.

25. Box, G.E.P. 1976. Science and statistics. *Journal of the American Statistical Association* 71: 791–799.

26. Bromham, L. 2016. *An Introduction to Molecular Evolution and Phylogenetics*. Oxford: Oxford University Press.

27. Bromham, L. 2019. Six impossible things before breakfast: Assumptions, models, and belief in molecular dating. *Trends in Ecology & Evolution* 34: 474–486.

28. Bromham, L., S. Duchéne, X. Hua, A.M. Ritchie, D.A. Duchéne, and S.Y.W. Ho. 2018. Bayesian molecular dating: opening up the black box. *Biological Reviews* 93: 1165–1191.

29. Brower, A. 2004. Comment on "molecular phylogenies link rates of evolution and speciation". *Science* 303: 173–173.

30. Bryson, V., and H.J. Vogel. 1965. Evolving genes and proteins. *Science* 68–71.

31. Carr, N. 2018. It's not a bug, it's a feature. *Wired* 26: 26.

32. Cheon, S., J. Zhang, and C. Park. 2020. Is phylotranscriptomics as reliable as phylogenomics? *Molecular Biology and Evolution* 37: 3672–3683.

33. Codani, J., J. Comet, J. Aude, E. Glãmet, A. Wozniak, J. Risler, A. Hãnaut, P. Slommski. 1999. 10 automatic analysis of large-scale pairwise alignments of protein sequences. In *Automation*, volume 28, ed. A.G. Craig, J.D. Hoheisel, 229–244 Cambridge: Academic Press. Methods in Microbiology.

34. Cole, D.B., D.B. Mills, D.H. Erwin, E.A. Sperling, S.M. Porter, C.T. Reinhard, and N.J. Planavsky. 2020. On the co-evolution of surface oxygen levels and animals. *Geobiology* 18: 260–281.

35. Cox, D.R. 2001. Comment on 'Statistical modeling: The two cultures'. *Statistical Science* 16: 216–218.

36. Cutler, D.J. 2000. Estimating divergence times in the presence of an overdispersed molecular clock. *Molecular Biology and Evolution* 17: 1647–1660.

37. Cuzzolin, F. 2021. *The Geometry of Uncertainty: The Geometry of Imprecise Probabilities*. Artificial Intelligence: Foundations, Theory, and Algorithms. Cham: Springer.

38. Dayhoff, M.O. 1972. *Atlas of Protein Sequence and Structure*. Washington: National Biomedical Research Foundation.

39. DeSalle, R., M. Tessler, J. Rosenfeld. 2020. *Phylogenomics: A Primer*. New York: CRC Press.

40. Donoghue, P.C., and M.P. Smith. 2003. *Telling the Evolutionary Time: Molecular Clocks and the Fossil Record*. New York: CRC Press.

41. Dos Reis, M., and Z. Yang. 2013. The unbearable uncertainty of Bayesian divergence time estimation. *Journal of Systematics and Evolution* 51: 30–43.

42. Drummond, A.J., and R.R. Bouckaert. 2015. *Bayesian Evolutionary Analysis with BEAST*. Cambridge: Cambridge University Press.

43. Ewens, W.J., and G.R. Grant. 2001. *Statistical Methods in Bioinformatics: An Introduction*. Statistics for Biology and Health. Berlin: Springer.

44. Fang, Z. 2021. The methods and tools for clustering analysis. In ed. O. Wolkenhauer. *Systems Medicine*, 9–13. Oxford: Academic Press.

45. Felsenstein, J. 1981. Evolutionary trees from DNA sequences: a maximum likelihood approach. *Journal of Molecular Evolution* 17: 368–376.

46. Felsenstein, J. 2004. *Inferring Phylogenies*. New York: Sinauer Associates.

47. Gillespie, J. 1994a. *The Causes of Molecular Evolution*. Oxford Series in Ecology and Evolution. Oxford: Oxford University Press.

48. Gillespie, J.H. 1984a. The molecular clock may be an episodic clock. *Proceedings of the National Academy of Sciences* 81: 8009–8013.

49. Gillespie, J.H. 1984b. Molecular evolution over the mutational landscape. *Evolution* 38: 1116–1129.

50. Gillespie, J.H. 1994b. Substitution processes in molecular evolution. II. exchangeable models from population genetics. *Evolution* 48: 1101–1113.

51. Gould, N.E.S.J., and N. Eldredge. 1972. Punctuated equilibria: An alternative to phyletic gradualism. *Essential Readings in Evolutionary Biology* 82–115.

52. Gould, S.J. 2009. *Punctuated Equilibrium*. Cambridge: Harvard University Press.

53. Hall, B. 2018a. *Phylogenetic Trees Made Easy: A How-To Manual*. New York: Sinauer Associates.

54. Hall, B. 2018b. Phylogenetic Trees Made Easy, Fifth Edition: 2018 Update for MEGA X. https://bit.ly/33r4fg9.

55. Harris, H. 1966. Enzyme polymorphisms in man. *Proceedings of the Royal Society of London. Series B, Containing papers of a Biological character. Royal Society (Great Britain)* 164: 298–310.

56. Hillis, D.M., and J.J. Bull. 1993. An empirical test of bootstrapping as a method for assessing confidence in phylogenetic analysis. *Systematic Biology* 42: 182–192.

57. Hoyle, F., and N. Wickramasinghe. 2000. *Astronomical Origins of Life*. New York: Springer.

58. Hu, T., M. Long, D. Yuan, Z. Zhu, Y. Huang, and S. Huang. 2013. The genetic equidistance result: misreading by the molecular clock and neutral theory and reinterpretation nearly half of a century later. *Science China Life Sciences* 56: 254–261.

59. Huang, S. 2012. Primate phylogeny: molecular evidence for a pongid clade excluding humans and a prosimian clade containing tarsiers. *Science China Life Sciences* 55: 709–725.

60. Huang, S. 2016. New thoughts on an old riddle: What determines genetic diversity within and between species? *Genomics* 108: 3–10.

61. Izadkhah, H. 2022. *Deep Learning in Bioinformatics: Techniques and Applications in Practice*. Amsterdam: Elsevier Science.

62. Jablonka, E., M. Lamb, and A. Zeligowski. 2014. *Evolution in Four Dimensions: Genetic, Epigenetic, Behavioral, and Symbolic Variation in the History of Life*. Cambridge: MIT Press.

63. Jeffreys, H. 1948. *Theory of Probability*. London: Oxford University Press.

64. Jukes, T.H., and C.R. Cantor. 1969. Evolution of protein molecules. *Mammalian Protein Metabolism* 3: 21–132.

65. Kaye, D., and J. Koehler. 2003. The misquantification of probative value. *Law and Human Behavior* 27: 645–659.

66. Kern, A.D., and M.W. Hahn. 2018. The Neutral Theory in Light of Natural Selection. *Molecular Biology and Evolution* 35: 1366–1371.

67. Kimura, M. 1968. Evolutionary rate at the molecular level. *Nature* 217: 624–626.

68. King, J.L., and T.H. Jukes. 1969. Non-Darwinian evolution. *Science* 164: 788–798.

69. Kishino, H., J.L. Thorne, and W.J. Bruno. 2001. Performance of a divergence time estimation method under a probabilistic model of rate evolution. *Molecular Biology and Evolution* 18: 352–361.

70. Koehler, J.J. 2002. When do courts think base rate statistics are relevant? *Jurimetrics* 373–402.

71. Koponen, I. 1995. Analytic approach to the problem of convergence of truncated lévy flights towards the gaussian stochastic process. *Physical Review E* 52: 1197–1199.

72. Kuhner, M.K., and J. Felsenstein. 1994. A simulation comparison of phylogeny algorithms under equal and unequal evolutionary rates. *Molecular Biology and Evolution* 11: 459–468.

73. Kumar, S. 2005. Molecular clocks: four decades of evolution. *Nature Reviews Genetics* 6: 654–662.

74. Kumar, S., G. Stecher, M. Li, C. Knyaz, and K. Tamura. 2018. MEGA X: Molecular evolutionary genetics analysis across computing platforms. *Molecular Biology and Evolution* 35: 1547.

75. Kumar, S., G. Stecher, M. Suleski, and S.B. Hedges. 2017. Timetree: A resource for timelines, timetrees, and divergence times. *Molecular Biology and Evolution* 34: 1812–1819.

76. Kumar, S., Q. Tao, S. Weaver, M. Sanderford, M.A. Caraballo-Ortiz, S. Sharma, S.L.K. Pond, and S. Miura. 2021. An evolutionary portrait of the progenitor SARS-CoV-2 and its dominant offshoots in COVID-19 pandemic. Molecular Biology and Evolution 38: 3046–3059.

77. Lage, K., E.O. Karlberg, Z.M. Størling, P.I. Ólason, A.G. Pedersen, O. Rigina, A.M. Hinsby, Z. Tümer, F. Pociot, N. Tommerup, Y. Moreau, S. Brunak. 2007. A human phenome-interactome network of protein complexes implicated in genetic disorders. *Nature Biotechnology* 25: 309–316.

78. de Laplace, M. 2009. *Essai philosophique sur les probabilités*. Cambridge: Cambridge University Press.

79. Lee, M.S.Y., and S.Y.W. Ho. 2016. Molecular clocks. *Current Biology* 26: R399–R402.

80. Lee, M.S.Y., J. Soubrier, G. Edgecombe. 2013. Rates of phenotypic and genomic evolution during the Cambrian explosion. *Current Biology* 23: 1889–1895.

81. Lee, W., K. Hopcraft, E. Jakeman. 2008. Continuous and discrete stable processes. *Physical Review E – Statistical, Nonlinear, and Soft Matter Physics* 77: 011109.

82. Lesk, A. 2019. *Introduction to Bioinformatics*. Oxford: Oxford University Press.

83. Lewontin, R.C., and J.L. Hubby. 1966. A molecular approach to the study of genic heterozygosity in natural populations. II. amount of variation and degree of heterozygosity in natural populations of drosophila pseudoobscura. *Genetics* 54: 595.

84. Lowen, S., and M. Teich. 2005. New York: John Wiley & Sons Ltd.

85. Manceau, M., J. Marin, H. Morlon, and A. Lambert. 2020. Model-based inference of punctuated molecular evolution. *Molecular Biology and Evolution* 37: 3308–3323.

86. Mandelbrot, B.B., and J.W. Van Ness. 1968. Fractional Brownian motions, fractional noises and applications. *SIAM Review* 10: 422–437.

87. Mantegna, R., and H. Stanley. 1994. Stochastic process with ultraslow convergence to a gaussian: The truncated lévy flight. *Physical Review Letters* 73: 2946–2949.

88. Margoliash, E. 1963. Primary structure and evolution of cytochrome C. *Proceedings of the National Academy of Sciences of the United States of America* 50: 672.

89. Mayo, D. 2018. *Statistical Inference as Severe Testing: How to Get Beyond the Statistics Wars*. Cambridge: Cambridge University Press.

90. Morgan, G.J. 1998. Emile Zuckerkandl, Linus Pauling, and the molecular evolutionary clock, 1959–1965. *Journal of the History of Biology* 31: 155–178.

91. Nakao, H. 2000. Multi-scaling properties of truncated lévy flights. *Physics Letters, Section A: General, Atomic and Solid State Physics* 266: 282–289.

92. Nei, M., and S. Kumar. 2000. *Molecular Evolution and Phylogenetics*. Oxford: Oxford University Press.

93. Nichol, S.T., J.E. Rowe, and W.M. Fitch. 1993. Punctuated equilibrium and positive Darwinian evolution in vesicular stomatitis virus. *Proceedings of the National Academy of Sciences* 90: 10424–10428.

94. Nutter, P.W. 2018. Machine learning evidence: admissibility and weight. *University of Pennsylvania Journal of Constitutional Law* 21: 919.
95. Ohta, T. 1973. Slightly deleterious mutant substitutions in evolution. *Nature* 246: 96–98.
96. Ohta, T. 1995. Synonymous and nonsynonymous substitutions in mammalian genes and the nearly neutral theory. *Journal of Molecular Evolution* 40: 56–63.
97. Ohta, T., and J.H. Gillespie. 1996. Development of neutral and nearly neutral theories. *Theoretical Population Biology* 49: 128–142.
98. Oswald, M. 2020. Technologies in the twilight zone: early lie detectors, machine learning and reformist legal realism. *International Review of Law, Computers & Technology* 34: 214–231.
99. Pagel, M., C. Venditti, A. Meade. 2006. Large punctuational contribution of speciation to evolutionary divergence at the molecular level. *Science* 314: 119–121.
100. Phillips, J.C. 2020. Self-organized networks: Darwinian evolution of dynein rings, stalks, and stalk heads. *Proceedings of the National Academy of Sciences* 117: 7799–7802.
101. Pipes, L., H. Wang, J.P. Huelsenbeck, and R. Nielsen, 2020. Assessing uncertainty in the rooting of the SARS-CoV-2 phylogeny. *Molecular Biology and Evolution* 38: 1537–1543.
102. Kliman, R.M. 2016. Phylogenetic tree comparison. In *Encyclopedia of Evolutionary Biology*, 277–284. Oxford: Academic Press.
103. Ritchie, A.M., X. Hua, M. Cardillo, K.J. Yaxley, R. Dinnage, and L. Bromham. 2021. Phylogenetic diversity metrics from molecular phylogenies: modelling expected degree of error under realistic rate variation. *Diversity and Distributions* 27: 164–178.
104. Salemi, M., A. Vandamme, and P. Lemey (eds.) 2009. *The Phylogenetic Handbook: A Practical Approach to Phylogenetic Analysis and Hypothesis Testing*. Cambridge: Cambridge University Press.
105. Schwabe, C. 2002. Genomic potential hypothesis of evolution: a concept of biogenesis in habitable spaces of the universe. *The Anatomical Record: An Official Publication of the American Association of Anatomists* 268: 171–179.
106. Schweizer, M. 2016. The civil standard of proof–what is it, actually? *The International Journal of Evidence & Proof* 20: 217–234.
107. Simpson, G.G. 1964. Organisms and molecules in evolution. *Science* 146: 1535–1538.
108. Stecher, G., K. Tamura, and S. Kumar. 2020. Molecular evolutionary genetics analysis (MEGA) for macOS. *Molecular Biology and Evolution* 37: 1237–1239.
109. Steele, E., R. Gorczynski, R. Lindley, Y. Liu, R. Temple, G. Tokoro, D. Wickramasinghe, and N. Wickramasinghe, 2019. Lamarck and panspermia – on the efficient spread of living systems throughout the cosmos. *Progress in Biophysics and Molecular Biology* 149: 10–32.
110. Takahata, N. 1987. On the overdispersed molecular clock. *Genetics* 116: 169–179.
111. Taleb, N. 2020. *Statistical Consequences of Fat Tails: Real World Preasymptotics, Epistemology, and Applications*. Austin: Scribe Media.
112. Taleb, N.N. 2007. *The Black Swan: The Impact of the Highly Improbable*. New York: Random House.
113. Tamura, K. 1992. Estimation of the number of nucleotide substitutions when there are strong transition-transversion and G+ C-content biases. Molecular Biology and Evolution 9: 678–687.
114. Tamura, K., Q. Tao, S. Kumar. 2018. Theoretical Foundation of the RelTime Method for Estimating Divergence Times from Variable Evolutionary Rates. *Molecular Biology and Evolution* 35: 1770–1782.
115. Tay, J.H., A.F. Porter, W. Wirth, and S. Duchene. 2022. The emergence of SARS-CoV-2 variants of concern is driven by acceleration of the substitution rate. *Molecular Biology and Evolution* 39.
116. Terdik, G., W. Woyczynski, and A. Piryatinska. 2006. Fractional- and integer-order moments, and multiscaling for smoothly truncated lévy flights. *Physics Letters, Section A: General, Atomic and Solid State Physics* 348: 94–109.
117. Torres, A., P.A. Goloboff, and S.A. Catalano. 2021. Assessing topological congruence among concatenation-based phylogenomic approaches in empirical datasets. *Molecular Phylogenetics and Evolution* 161: 107086.

118. Troffaes, M., and G. de Cooman. 2014. *Lower Previsions*. Wiley Series in Probability and Statistics. New York: Wiley.

119. Twain, M. 2004. *Life on the Mississippi* (accessed 14 August 2021). Urbana, Illinois: Project Gutenberg. https://www.gutenberg.org/ebooks/245.

120. Vaishnav, E.D., C.G. de Boer, J. Molinet, M. Yassour, L. Fan, X. Adiconis, D.A. Thompson, J.Z. Levin, F.A. Cubillos, and A. Regev. 2022. The evolution, evolvability and engineering of gene regulatory DNA. *Nature* 603: 455–463.

121. Waikagul, J., and U. Thaenkham. 2014. Methods of molecular study: DNA sequence and phylogenetic analyses. In ed. J. Waikagul, U. Thaenkham. *Approaches to Research on the Systematics of Fish-Borne Trematodes*, 77–90. Amsterdam: Academic Press.

122. Walley, P. 1991. *Statistical Reasoning with Imprecise Probabilities*. London: Chapman and Hall.

123. Walley, P. 2015. *BI Statistical Methods, Vol. 1: Foundations*. Dublin: Prescience Press.

124. Wang, M., D. Wang, J. Yu, and S. Huang. 2020. Enrichment in conservative amino acid changes among fixed and standing missense variations in slowly evolving proteins. *PeerJ* 8: e9983.

125. Webster, A.J., R.J. Payne, and M. Pagel. 2003. Molecular phylogenies link rates of evolution and speciation. *Science* 301: 478–478.

126. Weirich, P. 2021. *Rational Choice Using Imprecise Probabilities and Utilities*. Cambridge: Cambridge University Press.

127. West, B.J., and D.R. Bickel. 1998. Molecular evolution modeled as a fractal statistical process. *Physica A: Statistical Mechanics and Its Applications* 249: 544–552.

128. West, B.J., and D.R. Bickel. 1999. Fractional-difference stochastic model of evolutionary substitutions in DNA sequences. *Physics Letters, Section A: General, Atomic and Solid State Physics* 256: 188–196.

129. West, B.J., M. Bologna, and P. Grigolini. 2003. *Physics of Fractal Operators*. New York: Springer.

130. Wilson, A.C., S.S. Carlson, and T.J. White. 1977. Biochemical evolution. *Annual Review of Biochemistry* 46: 573–639.

131. Witt, C.C., and R.T. Brumfield. 2004. Comment on "molecular phylogenies link rates of evolution and speciation". *Science* 303: 173–173.

132. Xia, X. 2020. *A Mathematical Primer of Molecular Phylogenetics*. New York: Chapman and Hall/CRC.

133. Yang, Z. 2014. *Molecular Evolution: A Statistical Approach*. Oxford: Oxford University Press.

134. Yang, Z., and T. Zhu. 2018. Bayesian selection of misspecified models is overconfident and may cause spurious posterior probabilities for phylogenetic trees. *Proceedings of the National Academy of Sciences* 115: 1854–1859.

135. Zaslavsky, G. 2002. Chaos, fractional kinetics, and anomalous transport. *Physics Reports* 371: 461–580.

136. Zharkikh, A., and W.H. Li. 1992. Statistical properties of bootstrap estimation of phylogenetic variability from nucleotide sequences. I. Four taxa with a molecular clock. *Molecular Biology and Evolution* 9: 1119–1147.

137. Zuckerkandl, E., and L. Pauline. 1962. Molecular disease, evolution, and genetic heterogeneity. *Horizons in Biochemistry* 189–225.

138. Zuckerkandl, E., and L. Pauling. 1965. Evolutionary divergence and convergence in proteins. In *Evolving Genes and Proteins*, 97–166. Amsterdam: Elsevier.

Index

A
Alignment, 34–37, 41, 43, 47, 58–62, 64, 65, 67, 69, 70, 75, 81, 86
Autocorrelation, 92, 95, 96

B
Back substitution, 4, 33, 40
Bayesian, *see* Bayesian methods
Bayesian estimation, *see* Bayesian methods
Bayesian methods, 14, 49, 50, 63, 64, 66, 85
Bayes's theorem, *see* Bayesian methods
BEAST 2, 80, 81
Bioinformatics, 81–82, 86
BLAST, 36, 81
Body of evidence, 86–88
Bootstrap proportion, 39, 46, 51–53, 54, 81
    See also Frequentist methods

C
Calibration
    calibrated, 9
Common ancestor
    homologous, homology, 12, 33, 36, 79
Common descent, 12, 77
Confidence interval, viii, 14, 22, 37, 38, 46, 47, 49–55, 57, 76, 89
    See also Corrected confidence level; Frequentist methods
Continuous-stable distribution
    Lévy-stable distribution, 95
Continuous-stable process, 95–97
Corrected confidence level, 47, 51–54, 88, 89
    See also Confidence interval

Corrected tree, 33, 34, 41–43
Correction of unquantified uncertainty, 50–52, 66–67, 70, 80, 88–89
Counting process, 91, 92
Credible interval, *see* Bayesian methods

D
Democratic estimation, 59, 60, 68, 69
Discrete-stable process, 95, 96
Distance-based tree estimation, 29–32, 81
Distance matrix, 29, 30, 32–34, 41, 75
Divergence time, vii–ix, 6, 7, 9–11, 15–23, 45–55, 86, 87, 89

E
Empirical Bayes, 14, 20, 47, 64, 81
Epigenetic complexity, 72, 77
Estimating an ancestral DNA sequence, 58–60, 63, 64, 69
Evidential model, 86, 88, 89
Evidential sufficiency
    evidential sufficiency distribution, 87–88
Expected tree, 26–28, 34–37, 42, 60–62, 67–69

F
Fano factor, 92, 93
Fossil record
    fossils, 9, 16, 18–20, 22, 45, 48–50, 54, 72
Frequentist methods, 14, 19
    See also Confidence interval; Bootstrap proportion

Printed in the United States
by Baker & Taylor Publisher Services